Oxygen Radicals: Systemic Events and Disease Processes

Oxygen Radicals: Systemic Events and Disease Processes

Editors
Dipak K. Das, Farmington, Conn.
Walter B. Essman, Flushing, N.Y.

42 figures and 12 tables, 1990

KARGER

Basel · München · Paris · London · NewYork · New Delhi · Bangkok · Singapore · Tokyo · Sydney

Library of Congress Cataloging-in-Publication Data
Oxygen radicals: systemic events and disease processes/editors.
Dipak K. Das, Walter B. Essman.
p. cm.
Includes bibliographical references.
1. Active oxygen – Physiological effect. 2. Pathology, Cellular. I. Das, Dipak Kumar,
1946– . II. Essman, Walter B.
[DNLM: 1. Free Radicals. 2. Oxygen. QV 312 0967]
RB 170.0987 1990 616.07 – dc20 DNLM/DLC
ISBN 3–8055–5049–9

Drug Dosage
The authors and the publisher have exerted every effort to ensure that drug selection and
dosage set forth in this text are in accord with current recommendations and practice at the
time of publication. However, in view of ongoing research, changes in government regulations,
and the constant flow of information relating to drug therapy and drug reactions, the reader
is urged to check the package insert for each drug for any change in indications and dosage and
for added warnings and precautions. This is particularly important when the recommended
agent is a new and/or infrequently employed drug.

Contents

Contents VII

Contributors

Borrello, Sylvia, PhD,
Institute of General Pathology, Catholic University, Rome, Italy

Das, Dipak K., PhD,
University of Connecticut Medical Center, Farmington, Conn., USA

Engelman, Richard M., MD,
University of Connecticut Medical Center, Farmington, Conn., and
Baystate Medical Center, Springfield. Mass., USA

Essman, Walter B., MD, PhD,
Queens College, City University New York, Flushing, N.Y., USA

Forman, Henry J., PhD,
Childrens Hospital of Los Angeles, University of Southern California, Los Angeles,
Calif., USA

Galeotti, Tomaso, MD,
Institute of General Pathology, Catholic University, Rome, Italy

Jackson, Malcolm J., BSc, PhD,
Department of Medicine, University of Liverpool, Liverpool, England, UK

Masotti, Lanfranco, PHD,
Institute of Biological Chemistry, University of Parma, Parma, Italy

Rice-Evans, Catherine, PhD, BSc,
Royal Free Hospital School of Medicine, London, England, UK

Winterbourn, Catherine C., PhD,
Christchurch School of Medicine, Christchurch Hospital, Christchurch, New Zealand

Wollman, Stuart B., MD,
Department of Anesthesiology, Catholic Medical Center, Queens, New York, N.Y.,
USA

Foreword

In the year 2000 we will celebrate the centennial of Gomberg's discovery of free radicals – molecules that have missing atoms or groups. These unusual species with a ubiquitous unpaired electron defied the classical tenets of chemistry. It is interesting that this discovery involved stable free radicals. As a general rule, the majority of free radicals are extremely reactive and, consequently, short-lived species due to the presence of the unpaired electron. The technology in 1900, however, could not handle transient entities with very short lifetimes. It is remarkable, therefore, that Gomberg followed the only possible lead, which was the exception rather than the rule, namely, the stable free radicals. A relatively stable species, such as Gomberg's triphenylmethyl radical, is not found under normal physiological conditions, and the research of free radicals in biosystems had to await further developments.

In the 1930s Michaelis considered the formation of free radicals in bichemical reactions via one-electron transfer mechanisms, but again the proposition was premature. The major advances in free-radical chemistry that raised interest in bioradicals were brought about by the development of nuclear energy in the 1940s and 1950s. A great deal was learned from direct ESR measurements of 'frozen' radicals generated by radiation and the products of free-radical reactions in the aqueous phase under stationary-state conditions.

It is, however, the development of pulse radiolysis in 1962 by Hart, Boag, and Keene that marked the turning point in free radical research. Now the technology was available for kinetic detection and monitoring of short-lived radicals generated by submicrosecond pulses of high-energy (2 – 10 MeV) electrons. The next 4 years bristled with excitement because the rate constants of free-radical reactions, hitherto unaccessible, became available. Information about the reactivities and properties of hundreds of free radicals in model aqueous and organic systems paved the way for the discovery of the superoxide radical in biosystems by McCord and Fridowich in the late 1960s. This major discovery marked a turning point in free-radical research in biology and medicine.

This book deals with free-radical processes and their deleterious consequences in biosystems. It is clear now that human physiology is equipped to deal with free radicals and prevent their damaging effects. Often, however, the protective mechanisms are breached, and the living organism suffers from the resulting consequences. In addition to the damage to the genetic material, there is hardly any organ that is spared. It is hoped that the review articles in this book, which summarize the endeavors of researchers in this fascinating field, will provide an educational base and stimulate further and faster the advances needed not only to understand but also to control free-radical processes *in vivo* for the benefit of all of us.

Michael G. Simic, PhD, DSc
National Bureau of Standards

Preface

Free radicals (molecules or atoms with an unpaired electron) and their metabolites have been known for their effect upon mechanisms regulating cellular function. Several metabolic events can account for free radical formation, e.g. lipid metabolism. Molecular oxygen can act as an electron acceptor, by which several cellular free radical reactions occur. In the mitochondrial electron transport system molecular oxygen serves as the final electron acceptor.

There are numerous tissue processes associated with inflammation, ischemic injury, drug effects, etc., that are mediated by oxygen-derived free radicals.

The product of the acceptance of an electron from a reducing agent by molecular oxygen is the superoxide anion. Superoxide is in equilibrium with its protonated form, $HO_2^{\cdot}$ in an aqueous environment. At a neutral pH, superoxide is the dominant species, as the pK_a for the equilibrium reaction is 4.8. In acidic environments relatively high $HO_2^{\cdot}$ concentrations are favored. Organic molecules are capable of being oxidized by several superoxide derivatives including the $HO_2^{\cdot}$ radical. Spontaneous dismutation will result when superoxide and the protonated form of the superoxide anion begin to occur in equimolar concentration. As a dismutation product, H_2O_2 is generated; it may also occur from a double reduction of molecular oxygen, or from enzyme-catalyzed dismutation of superoxide. Products of the dismutation of two superoxide molecules by the enzyme superoxide dismutase (SOD), are hydrogen peroxide and oxygen. This reaction occurs with a rate constant approximately four times that of spontaneous superoxide dismutation.

Cellular access to oxygen-derived free radicals occurs either by direct cell membrane crossing by H_2O_2 or via membrane channels in the case of superoxide radical. There are other free radicals derived from molecular oxygen that have also been implicated in cellular and tissue injury; these include: (1) Singlet oxygen, which is formed when one of the two unpaired electrons of molecular oxygen absorbs sufficient energy to create spin inversion and orbital transition. Several reactions may provide a basis for singlet oxygen formation, such as direct radiation absorption; singlet oxygen is capable of reacting with electron-rich compounds, such as tryptophan, methionine, and histidine – primarily based upon its strong electrophylic property. (2) The hydroxyl radical, a product of the Haber-Weiss reaction, the Fenton reaction, or through an iron-H_2O_2-iodine system. (3) H_2O_2 reaction with peroxidase in the presence of halides. Hypohalous acids can form when the reaction of H_2O_2, in the presence of halides, occurs with a variety of organelle-specific peroxidases. Some of these oxidant acid products have bacteriacidal activity.

Oxygen-derived free radicals can occur in a variety of tissues, exerting myriad effects, and capable of interacting with a range of pharmacological agents possessing anti-oxidant properties for several systems. The focus of this monograph is the nature of oxygen-derived free radical activities for several physiological systems. The significance of such activities relates to an understanding of the biological properties of free-radical actions, as well as to the importance of free-radical effects for medicine.

Walter B. Essman
Dipak K. Das

1 Erythrocytes, Oxygen Radicals and Cellular Pathology

Catherine Rice-Evans

Oxygen-derived free radical damage is associated with shortened red cell survival in a number of haemolytic states due either to congenital or acquired defects, such as oxidant drug-induced haemolysis [1] and thalassaemia major [2, 3] as well as in the aging of erythrocytes [4, 5]. The production of active oxygen species and the susceptibility of oxyhaemoglobin to autoxidation relates to the ability of an electron to become depolarised from the haem iron to the bound oxygen which, on deoxygenation is normally returned to the iron when the oxygen is released. The globin-moiety and, specifically, the non-polar niche that contains the haem, is indispensable for the maintenance of iron in the ferrous form that is essential for its ability to bind oxygen reversibly [6]. Any slight modification in the haem pocket which may allow access of small anions or water molecules displaces the electron from the haem iron, resulting in the formation of the superoxide radical [7] and methaemoglobin [8].

In the normal course of events there is a small degree of spontaneous autoxidation of haemoglobin [9] with the production of 10^7 superoxide radicals per day [7, 10] within the erythrocyte. This is amply counteracted in the normal state by the antioxidant defences of the cell. Any pathological situation which increases the turnover of this cycle, whether increased oxidative stress or impaired antioxidant defences, will enhance the production of active oxygen species. In this chapter, pathological situations in which the antioxidant defences are unable to compensate for the ongoing oxidative stress will be described. Three such disorders will be portrayed: (1) sickle cell anaemia and the implications of membrane lipid hydroperoxides, (2) thalassaemia major in which excess iron is involved extra cellularly, intracellularly in the form of membrane-bound haemichrome, and possibly in the form of intracellular iron complexes residual from the thalassaemic reticulocyte, and (3) glucose-6-phosphate dehydrogenase (G6PD) deficiency in which the detoxification mechanisms for hydrogen peroxide are deficient.

Red Cell Survival

The function and survival of the human erythrocyte is dependent on its ability to remain flexible, durable and deformable enough to pass through the capillaries and the narrow portals of the spleen as it proceeds on its 120-day journey through the circulation [11]. During this time it will traverse the circulation half a million times covering a distance of 300 miles [12]. The tiny slits between the splenic cords appear very constraining and only red cells that are extremely deformable and of unswollen dimensions escape entrapment. The ability of the red cell to deform or change its shape in response to an applied force is thus vital to its survival. Consequently, it is of great importance in the understanding of the pathophysiology of erythrocyte disorders in which reduced survival is a prominent characteristic.

The relative ability of the red cell to retain its deformability is dependent on: the maintenace of its biconcave discocytic morphology, the low viscosity of the intracellular contents, the energy state of the cell and the viscoelastic properties of the membrane, all of which are inter-related [13, 14]. The erythrocyte in its discocytic morphology has an advantageous surface area to volume relationship, allowing it to undergo a marked deformation while maintaining a constant surface area; if the cell loses membrane due to fragmentation while maintaining its volume (i.e. loss of surface area) then the cell becomes spherocytic and loses this advantage, leading to a loss of deformability. Under normal conditions the low viscosity and fluid nature of the haemoglobin within the red cell do not impose any constraints on the ability of the red cell to deform, unless the cells become dehydrated or shrunken, causing an increase in the mean cell haemoglobin concentration and increased internal viscosity.

Pathological cells that are dehydrated and those containing precipitated or polymerised haemoglobin would have decreased deformability, e.g. thalassaemic or sickle cell erythrocytes. Erythrocytes with damaged or modified membranes become less deformable and may be subject to mechanical entrapment in the microcirculation, where they are ultimately recognised and sequestered by phagocytic cells. In recent years the importance of the membrane skeletal proteins has become well-recognised in contributing towards the stability and the viscoelastic properties of the erythrocytes, accounting for their remarkable deformability and capacity to survive in the turbulent bloodstream [15].

Erythrocyte Membrane Structure

The human erythrocyte membrane consists of a bilayer of phospholipids arranged asymmetrically, intercalated within which are the intra-

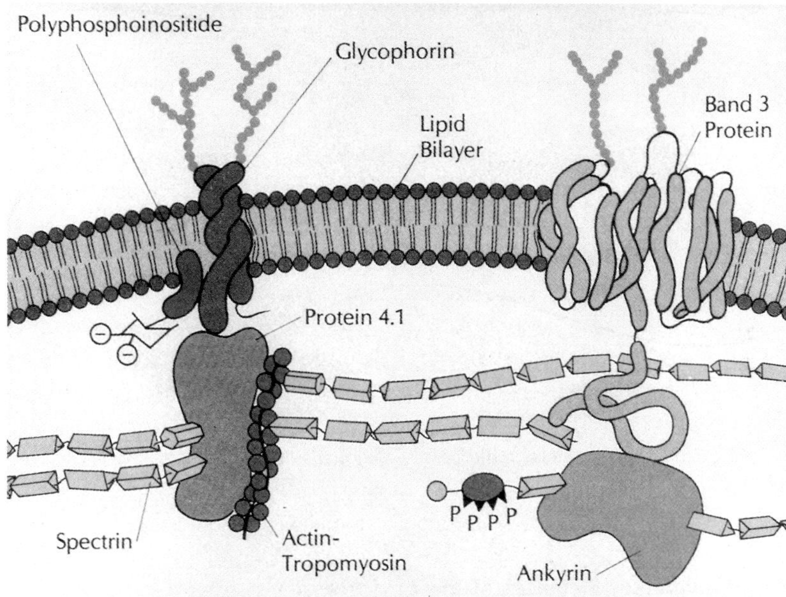

Fig. 1. The arrangement of the membrane proteins of the human erythrocyte [15].

membranous glycoproteins and glycolipids. These carry the carbohydrate
components, bearing the determinants of recognition and specificity on the
outer surface of the cell. The membrane skeleton is the critical intermediary
at the interface between the cell membrane and the cytoplasmic constituents
and it is now known to perform several functions. This protein framework
is important in the maintenance of cellular shape and the viscoelastic properties of the membrane, the control of phospholipid asymmetry and stability,
the transmission of signals from the external environment to the cell interior
and the modulation of protein movement within the membrane. The membrane skeleton (fig. 1) [15] comprises three proteins: the predominant one,
spectrin, forms fibres in the form of tetramers, each consisting of two dimers
joined head-to-head. The ends of each tetramer are linked to the next by much
shorter fibres of actin. The spectrin-actin linkage is mediated by protein 4.1
which binds to the tails of the spectrin fibres near the actin binding site. The
cytoskeletal network is attached to the membrane via ankyrin which is
linked to the spectrin tetramer near its centre. Ankyrin itself binds to the
intramembranous protein 3, the major integral membrane protein, the anion
transporter. The other major membrane-spanning protein is the sialoglyco-
protein, glycophorin, which is linked to the cytoskeletal network through
protein 4.1, but only in the presence of polyphosphinositides [15].

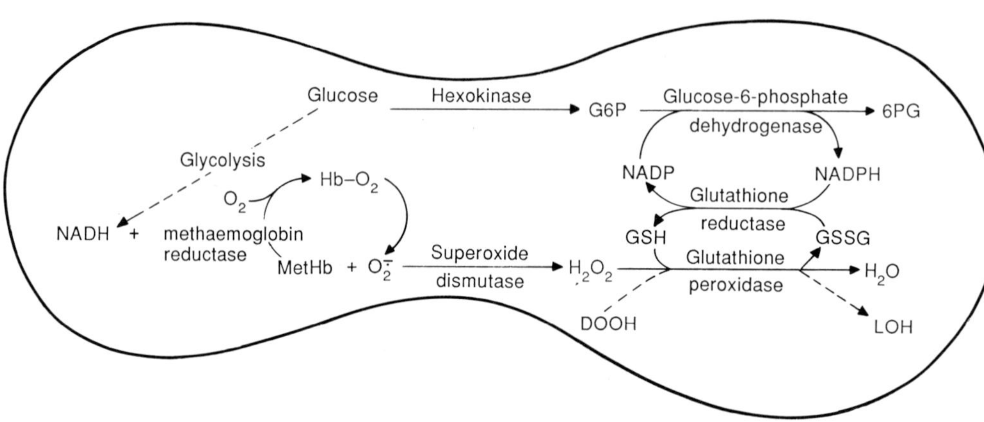

Fig. 2. The intracellular antioxidant defences of the human erythrocyte.

Antioxidant Defences of the Erythrocyte

The erythrocyte is well-endowed with a number of interacting enzymatic and non-enzymatic systems that serve to modulate the reactivity of free radicals. Some of these systems are located in specific compartments, in the cell membrane, in the cytoplasm and extracellularly (fig. 2). On autoxidation of haemoglobin, NADH-dependent methaemoglobin reductase converts the methaemoglobin back to deoxyhaemoglobin. Cu,Zn-superoxide dismutase handles the superoxide [16] formed from haemoglobin oxidation and dismutates it to hydrogen peroxide [16, 17] which has limited reactivity; reduced glutathione and glutathione peroxidase detoxify the hydrogen peroxide [18]; however, maintenance of reduced glutathione levels is essential for the sustaining of the thiol groups of the cellular enzymes, the intramembranous and cytoskeletal proteins in the reduced state and the cell is equipped with NADPH-dependent glutathione reductase to re-convert the oxidised glutathione to the reduced state. Tocopherol is one of the most important antioxidants residing mainly in the membrane where it can function as a chain-breaking antioxidant, interrupting the propagation of oxygen radical mechanisms [19, 20]. Major plasma antioxidant mechanisms include ascorbate, which works synergistically with tocopherol, transferrin which binds iron (III) for transport and caeruloplasmin which acts as a ferroxidase. It has been suggested that caeruloplasmin may function in facilitating iron binding to transferrin [21] and perhaps, in the case of iron overload, to plasma ferritin [22].

The Role of Iron in Oxygen Radical Production

The major role of superoxide radicals in oxygen toxicity is as a reducing agent for metal ions:

$$Fe^{3+} + O_2^{\cdot-} \rightarrow Fe^{2+} + O_2$$

Other reducing agents are also present which may perform similar functions, e.g. ascorbate [23–25], reduced glutathione and other thiols [26–27], reduced nicotinamide complexes NADH, NADPH [28]. The reduction of iron (III) via this reaction accompanied by the dismutation of superoxide radical to hydrogen peroxide can predispose the cell to the deleterious effects of the extremely reactive hydroxyl radical via the Fenton reaction [29]:

$$Fe^{2+} + H_2O_2 \rightarrow Fe^{3+} + OH^- + OH^{\cdot}$$

The combination of these two reactions gives the iron-catalysed Haber-Weiss reaction [30]:

$$O_2^{\cdot-} + H_2O_2 \rightarrow OH^{\cdot} + OH^- + O_2$$

But are redox metals available at the site where superoxide radical is formed or does the superoxide move through the cell? It has been reported that much of the superoxide generated within cells in general comes form membrane-bound systems [31]. The hydroxyl radical will react with any species in its proximity, its radius of interaction being less than 100 nm [32], and will initiate lipid peroxidation, damage proteins, attack carbohydrates and oxidise intracellular components, depending on its site of production. The major question which then arises is whether iron catalysts are available for the Fenton and Haber-Weiss reactions in vivo. Caeruloplasmin, by removing iron (II) via its ferroxidase action, is an important extracellular defence system in vivo assisting in the removal of iron (II) and thereby preventing hydroxyl radical formation: in maintaining iron ions in the oxidised state it is able to prevent the initiation of lipid peroxidation. Similarly, transferrin is an important extracellular antioxidant since this protein is only partially iron-loaded in vivo, except in thalassaemia (see later) and therefore sequesters iron, making it unavailable for either catalysing the Haber-Weiss reaction or for initiating peroxidation by interaction with molecular oxygen or for performing the decomposition of lipid hydroperoxides. Furthrmore, no free iron pool has been observed in normal blood serum [33] or within the normal erythrocyte.

Oxidative Damage in Erythrocytes

Numerous functions of erythrocytes and their membranes are perturbed when under oxidative stress, whether drug-induced or in a pathological situation of either an unstable haemoglobin or in which the antioxidant defences are impeded, eventually leading to cell lysis. Red cell damage by oxidant stress is generally thought to be the end result of three processes: (1) the oxidation of haemoglobin followed by the denaturation of methaemoglobin to haemichromes; (2) free radical attack on the polyunsaturated fatty acids of the membrane lipids and (3) radical attack on specific amino acids and reduced thiol groups of the membrane proteins and other susceptible components.

Haemoglobin Oxidation

Oxidation of haemoglobin is followed by the denaturation of methaemoglobin to haemichromes (fig. 3). Haemichromes represent oxidation states of altered iron-porphyin complexes [34]. Initially, reversible haemichromes are formed, which are capable of reduction to functional haemoglobin, distortion of the haem pocket allowing the formation of an internal sixth ligand from the globin moiety to iron results in the formation of irreversible haemichromes which denature and precipitate. Ultimately the haem dissociates and Heinz body formation occurs [35], these binding to the membrane to form inclusion bodies, eventually leading to cell lysis.

In unstable haemoglobins, such as the thalassaemias, in which one of the globin chains is deficient or absent, the chains are directly oxidised to the haemichrome state, prior to the formation of Heinz bodies via the oxidation of isolated chains [36]. This indicates that the association of α- and β-globin chains to form tetramers in haemoglobin acts as a protective mechanism against the attack of oxidants present in red cells, such as hydrogen peroxide or superoxide, a protection which is defective in diseases such as thalassaemia.

The Membrane Lipids and Reactive Oxygen Species

The peroxidation of polyunsaturated fatty acid side chains of the membrane lipids is a consequence of many types of cellular injury in which free radical intermediates are produced in excess of local defence been mechanisms and has been extensively reviewed [32, 37–39].

Lipid peroxidation may be initiated by any primary reactive free radical species. The production of a highly reactive free radical leads to primary reactions and damage in the immediate vicinity of where the radical is produced–it will not diffuse far before it interacts within its

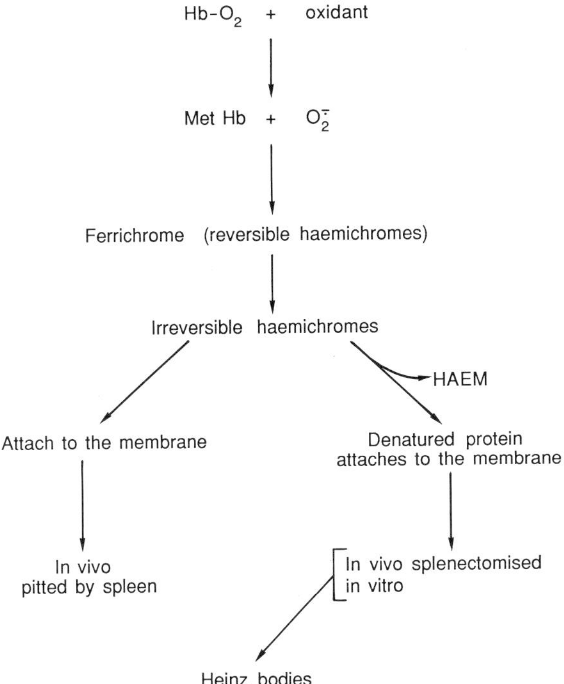

Fig. 3. The oxidative denaturation of haemoglobin.

microenvironment which in a membrane would include initiation of lipid peroxidation. Secondary products of lipid peroxidation such as lipid peroxy radicals and lipid hydroperoxides may diffuse in the plane of the membrane before reacting further, thereby spreading the biochemical lesion. The consequences of such processes are twofold: firstly, they affect the structural and functional integrity of the membrances, including altered lipid fluidity and modified interactions with neighbouring components, and secondly the breakdown products of lipid peroxidation can further damage cellular function. For example, alkenals, alkanals, lipid hydroperoxides may undergo degradative reactions and be metabolised rapidly; some of these like the lower molecular weight hydroperoxides, aldehydes and 4-hydroxyalkenals can escape from the membrane and cause disturbances at a distance. Therefore a reaction that originally produces a radical which interacts within its own microenvironment, may produce a sequence of later events that direct disturbances throughout the cell, its membrane and in some instances the extracellular domain.

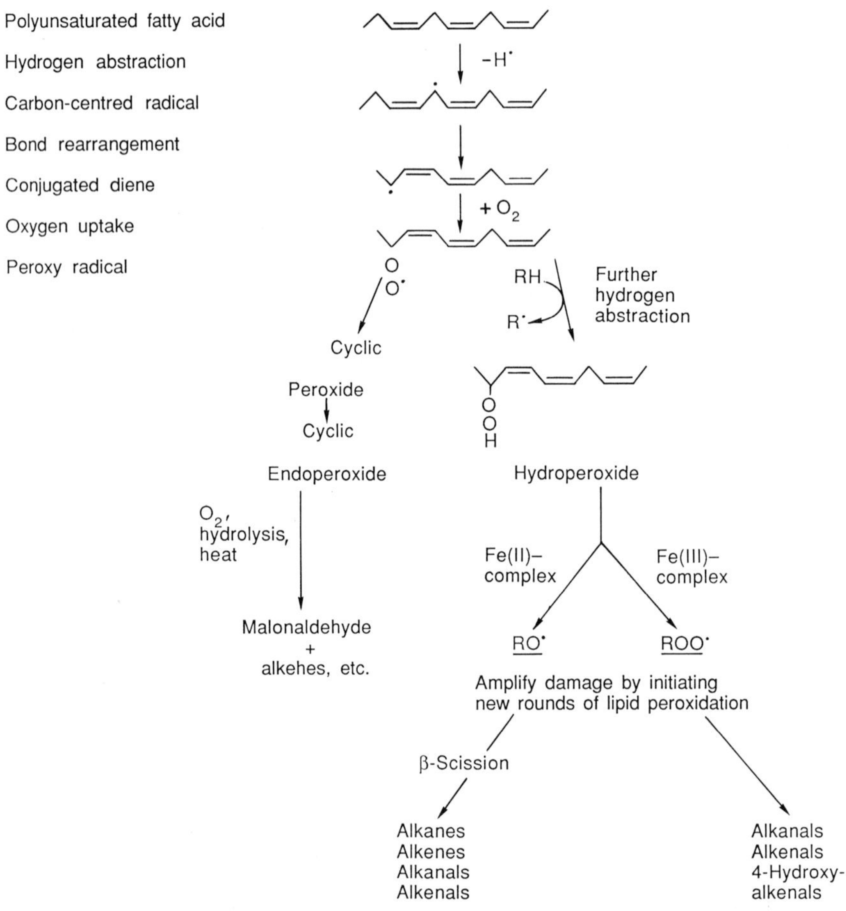

Fig. 4. The mechanism of the peroxidative degradation of a polyunsaturated fatty acid side-chain.

The mechanism of peroxidation of a polyunsaturated fatty acid side chain of a membrane lipid is depicted in figure 4. Any primary reactive free radical species which has sufficient reactivity to abstract a hydrogen atom from a reactive methylene group of an unsaturated fatty acid side chain can initiate lipid peroxidation. This would include such species as the hydroxyl radical, ·OH, alkoxy radical, ·OR, peroxy radical, ·OOR, alkyl radical, R·, etc. The formation of the initiating species is accompanied by bond rearrangement that results in stabilisation by diene conjugate formation. The lipid radical then takes up oxygen to form the peroxy radical.

Propagation reactions follow, leading to the formation of lipid monohydroperoxides. This propagation phase can be repeated many times—thus an initial event triggering lipid peroxidation can be amplified as long as oxygen supplies and unoxidised fatty acid chains are available, which is exactly the case in the red cell.

Lipid hydroperoxides are fairly stable molecules at physiological temperatures but their decomposition is catalysed by transition metals and metal complexes [40]. Both iron (II) and iron (III) are effective catalysts for hydroperoxide degradation but the latter more so [41], e.g. complexes of iron salts with phosphate compounds, non-haem iron proteins, free haem, haemoglobin, ferritin (to an extent proportional to the degree of iron loading):

$$Fe^{2+} + LOOH \rightarrow Fe^{3+} + LO^{\cdot} + OH^{-}$$

$$Fe^{3+} + LOOH \rightarrow Fe^{2+} + LOO^{\cdot} + H^{+}$$

Alkoxy and peroxy radicals can initiate new rounds of lipid peroxidation and propagate further radical chain reactions thus amplifying the initial damage. Cleavage of carbon-carbon bonds during such peroxidative reactions results in the formation of alkanals such as malonaldehyde [42] or alkenals, such as 4-hydroxynonenal, which is very biologically active: it can inhibit platelet aggregation, adenyl cyclase activity [43–45]. Alkyl radicals are also formed during the cleavage of carbon-carbon bonds which can initiate new rounds of lipid peroxidation-forming alkanes: pentane is produced by this mechanism as an end-product of the oxidation of linoleic and arachidonic acid and ethane from linolenic acid [42, 46, 47].

Radical Effects on Membrane Proteins
Proteins may be critical targets of free radical attack as they are present both inside and outside cells in very high concentrations, the outcome being the accumulation of denatured protein which may interfere with cellular function. Several amino acids crucial for protein function are particularly susceptible to radical damage, for example, methionine, tryptophan, histidine, cysteine and tyrosine. Radical-mediated protein damage can arise from either a direct attack of radical species, e.g. hydroxyl radicals generated from hydrogen peroxide accumulation in the presence of low molecular weight chelates of iron or copper, or as a consequence of the formation of radical intemediates of lipid peroxidation: lipid hydroperoxides are stable intermediates of peroxidation which can react with transition metals generating alkoxy and peroxy radicals, as described previously. These may react with proteins closely associated with the peroxidising lipids causing fragmentation or cross-linking [48]. Furthermore, studies have

shown that two of the stable products of lipid peroxidation, namely malonaldehyde and 4-hydroxynonenal, can modify lysine residues [49] for example.

The consequences for the cell may be: (1) altered enzymic activities–in the main, enzyme inactivation would occur, but it is possible that enzyme activation might result where an enzyme inhibitor becomes inactivated; (2) increased susceptibility to proteolytic attack and enzymic hydrolysis [48], for example, haemoglobin becomes more susceptible to enzymic hydrolysis after exposure to hydrogen peroxide; (3) altered membrane and cellular function due to cross-linked proteins via disulphide oxidations, dityrosine links, etc. [49], or degradation of functional proteins such as the membrane transport proteins, leading to modified ionic effects in the cell.

Lipid Hydroperoxides in the Pathology of Sickle Cell Anaemia

Several groups have now presented evidence that sickle erythrocytes are under oxidative stress [50–52] and that this is amplified by the binding of degradation products of haemoglobin to the membrane, which may catalyse the breakdown of lipid hydroperoxides. The mechanism of the initiation of such oxidised species in the membrane and the relationship with the sickling process, during which the binding of deoxyhaemoglobin to the membrane occurs, are far from clear.

The primary defect in sickle cell anaemia is the increased viscosity of the deoxygenated blood due to the polymerisation of the intracellular haemoglobin which distorts the cells into drepanocytes (fig. 5). On reoxygenation the normal discocytic morphology is restored in the first instance. During the process of reversible sickling a poorly defined membrane injury seems to be induced, that may be due to membrane distortion by the binding of haemoglobin S, culminating in the formation of irreversibly sickled cells (ISC). These cells remain sickled and fail to function normally even when the haemoglobin is completely reoxygenated [53]. This implies that the membrane is an important locus of secondary cellular damage and is most interestingly manifested in the modified shape of the cytoskeletal protein network [54]: irreversibly sickled cells have a sickle-shaped membrane, whereas reversibly sicklable cells have normal spherocytic membranes; separation of the lipids and the intramembranous proteins reveals a sickle-shaped cytoskeleton [55] from the ISC. During repeated sickling cycles the cytoskeletal proteins have undergone structural modifications. Hence, what started out as a genetic defect in the β-globin chains of the haemoglobin structure has now been amplified into a gross membrane abnormality.

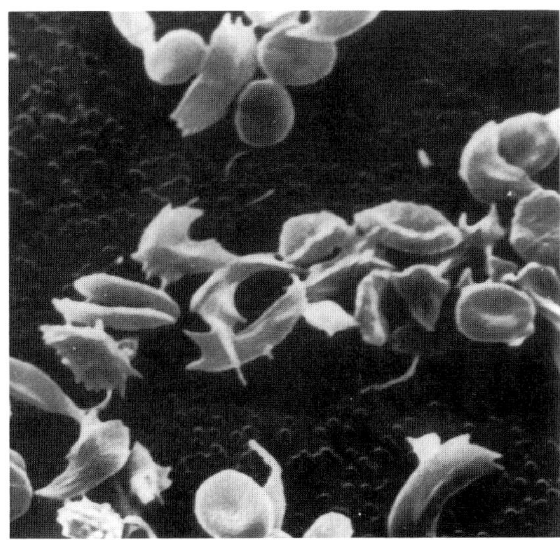

Fig. 5. The drepanocytic morphology of the sickled erythrocyte.

Sickle cells become dense and dehydrated in the circulation generating a variable fraction of cells with increased mean cell haemoglobin concentration. These cells have decreased deformability [56, 57] which is accentuated by the increased polymerisation of sickle haemoglobin on partial deoxygenation. A major question here is the influence of membrane-bound haemoglobin on the altered membrane characteristics of sickle cell membranes. During the sickling process the cells undergo increased membrane permeability to cations, especially calcium ions [58] and potassium ions with a corresponding loss of cell water [59], altered phospholipid asymmetry [60], decreased lipid fluidity [61], the formation of a defective cytoskeleton [54] (as described above), altered membrane surface charge [61], increased adherence to vascular endothelium [62] and enhanced susceptibility to oxidative stress [50–52].

Altered Permeability to Cations
Sickle cells contain elevated levels of calcium ions [58] and decreased adenosine triphosphate (ATP) levels [61]. Although levels of calcium attained by circulating sickle cells are high enough ($100-300\ \mu M$) [58] to activate endogenous transglutaminases, high molecular weight protein aggregates have not been detected. The accumulation of calcium ions may activate endogenous proteases thus causing damage to ankyrin or spectrin or neighbouring proteins [63] by excising critical peptides and modifying cytoskeletal integrity. The implications of modified intracellular calcium levels for cytoskeletal protein interactions and organisation of membrane

lipids are still far from clear. However, if increased oxidative stress is a cause rather than a consequence of increased cation permeability, then any oxidative-modified proteins would become more susceptible to the effects of calcium-activated proteases as described previously.

It has also been suggested that increased calcium levels have a profound effect on phospholipid asymmetry [64] whereas the studies of Bookchin et al. [65] have indicated that the increased calcium is sequestered in vesicles within the sickle erythrocyte and as such would be less available for interaction with the cytoskeletal proteins or the membrane.

Modified Phospholipid Asymmetry

Membrane lipid reorganisation has been shown to occur during the sickling process. Increased amounts of phosphatidyl serine and phosphatidyl ethanolamine have been located on the outer layer of the cell membrane suggesting a disruption in the stabilisation and organisation of the phospholipid bilayer [59]. Since phospholipid asymmetry is said to be stabilised by interactions with spectrin and neighbouring proteins, this may be a direct consequence of a modified cytoskeleton during the sickling process.

Vesiculation

Another interesting property of sickle cells is the tendency of the membrane to vesiculate during the sickling process. It has been shown that sickle erythrocytes which are repeatedly deoxygenated and reoxygenated in vivo lose 2–3% of their lipids in the form of long spicules and small spheres [66, 67]. Eventually the long spicules exocytose with the formation of microvesicles. These can be observed morphologically in samples from sickle cell patients. The vesicles have been shown to contain haemoglobin, glyceraldehyde-3-phosphate dehydrogenase, proteins 3 and 4.1, but are deficient in spectrin [66–68]. The phospholipid composition is the same as that in the sickle cell membrane.

Allan et al. [66–67] have proposed that as cells sickle filaments of polymerised haemoglobin protrude through and disrupt the cytoskeletal protein network, breaking its interaction with the membrane lipid bilayer. Altered distribution of the phospholipids is said to result which may facilitate the formation of the long spicules erythrocyte which eventually vesiculate. However, the process is clearly more complex than this since vesiculation of the sickle erythrocyte has been observed without sickling [69]. More recent work indicates that destablisation of the lipid bilayer in sickle cells is manifested in enhanced flip-flop of phosphatidyl choline and that exposure of phosphatidyl serine in the outer monolayer occurs predominantly in those areas of the membrane that are in spicular form [70].

Abnormal Cellular Interactions

Sickle erythrocytes have an abnormal propensity for adherence to and phagocytosis by macrophages which may contribute towards the accelerated destruction of the erythrocytes in sickle cell disease [71]. It has also been suggested that sickle cells are abnormally adherent to vascular endothelial cells and that aberration in surface charge topography may be responsible [62]. It is possible that the reported changes in the orientation of the membrane lipids [60] and in charge at the membrane surface [61] play a role in the adhesive properties of the sickle cell membrane.

Several groups have presented evidence for oxygen radical formation in sickle erythrocytes [50–52]. Rice-Evans et al. [50, 72] have suggested that the membranes of fresh sickle erythrocytes contain endogenous products of lipid peroxidation. Exposure of sickle cells to exogenous oxidative stress in the form of added hydrogen peroxide has also been reported to result in the accumulation of lipid breakdown products [73, 74] and that after a 20-hour incubation sickle erythrocytes generate twice as much superoxide, peroxide and hydroxyl radical [52]. Furthermore, in sickle cell anaemia the reductive mechanisms of the cell against oxidative damage are defective with decreased activities of gluthathione peroxidase and catalase [73], increased levels of superoxide dismutase [73] and decreased levels of vitamin E in plasma and membrane [74]. Similar modifications have been reported in other blood disorders in the pathology of which oxidative mechanisms have also been implicated, for example, thalassaemia major and G6PD deficiency.

Erythrocytes from patients with sickle cell anaemia [50, 75, 76] can be divided into two groups according to their pathological state in relation to the proportion of ISC and to their oxidative status in terms of their content of lipid peroxidation products: one group of samples with a higher proportion of ISC (5–25%) contain elevated endogenous levels of secondary products of oxidative damage, resulting in pronounced modifications in the membrane properties, the other group of samples with a lower proportion of ISC (< 5%) and with apparently more normal characteristics (table 1). This classification is consistent with earlier observations [61] demonstrating a correlation between the proportion of ISC and the degree of haemoglobin retention in the sickled erythrocyte with altered membrane properties.

The presence of endogenous products of lipid peroxidation in sickle cell membranes and the increased accumulation of lipid breakdown products after oxidative stress induced by incubation implicate membrane-bound denaturation products of haemoglobin as propagators of such damage. Furthermore, exposure of sickle cells to exogenous oxidative stress generates increased production of superoxide and hydroxyl radicals [52]. The interpretation is that hydroxyl radical generation may be facilitated by membrane-bound haemichrome catalysed reactions between the superoxide

Table 1. Indication of endogenous oxidative stress in sickle erythrocytes: assessment of metabolites of lipid peroxidation and thiol oxidation status of membrane proteins

	Sickle erythrocytes		Normal erythrocytes
	> 5% ISCs	< 5% ISCs	
Lipid peroxidation (thiobarbituric acid-reactive products)			
Erythrocytes ($nM/10^{10}$ cells)	11.20 ± 2.6 (18 patients)	6.00 ± 1.4	6.95 ± 2.5 (10)
Membranes A_{332}/mg protein	0.15 ± 0.3 (24 patients)	0.08 ± 0.04	0.072 ± 0.01 (21)
Protein thiol oxidation (nM SH/mg protein)	65 ± 5 (24 patients)	80 ± 3	84 ± 9 (24)

radical and hydrogen peroxide in the same way that traces of transition metals have been shown to catalyse the Haber-Weiss reaction [30]. But a criticism of this interpretation is that native haemoglobin, haem, methaemoglobin, etc. will not catalyse hydroxyl radical formation production via these mechanisms an open coordination site on the iron ion or a site only loosely associated with ligands is necessary [78]. However, iron in the form of haem, haemoglobin, methaemoglobin and other components has a role in the decomposition of lipid hydroperoxides to alkoxy and peroxy radicals [40].

It has been proposed that the initial event in the sickling process is the irreversible binding of haemoglobin S to the membrane [79]. Normally in the metabolism of the erythrocyte there exists the capacity for the restoration of any oxidised haemoglobin to its normal functional state. However, in situations in which autoxidation may be followed by the formation of haemiochromes, for example, reduction mechanisms become overwhelmed and cannot counteract efficiently this conversion leading to the accumulation of oxidised products of haemoglobin.

Experiments on the extent of such binding by measuring the amount of iron bound to the sickle cell membrane compared with that in normal haemoglobin-free membranes prepared under the same conditions have been carried out using atomic absorption spectroscopy [80] and by measuring haemichrome formation [81]. The results correspond with about 0.7% of the total haemoglobin in sickle erythrocytes bound to the membrane, within this group of patients, about 50–80 times more than in normal erythrocyte membranes.

The possible role of iron, in the form of membrane-bound denaturation products of haemoglobin, in the propagation of oxidative damage in sickle erythrocytes has been shown from experiments on isolated sickle erythrocyte membranes. Sickle cell membranes incubated for 5 h at 37 °C with no additives generated elevated levels of products of lipid peroxidation in the form of thiobarbituric acid-reactive compounds [50] whereas normal erythrocyte membranes were unmodified in this respect. Similar results were obtained when sickle erythrocytes were stressed by incubation under the same conditions [50]. The addition of 5 mM ascorbate in this system prevented the increase in lipid peroxidation which would have been observed in the sickle cell membranes in its absence. Ascorbate may be acting here in its capacity as a scavenger of alkoxy or peroxy radicals by a chain termination reaction [41] or by decreased initiation of lipid peroxidation. Incorporation of desferrioxamine, the iron chelator applied therapeutically in the treatment of thalassaemia major, also significantly depressed the lipid peroxidation below the level observed pre-incubation. Ferric ion chelated to desferrioxamine is not free to catalyse the decomposition of lipid hydroperoxides nor to participate in other potential iron-mediated reactions [82]. However, iron in haem, haemichrome, methaemoglobin, etc. is not normally chelatable by desferrioxamine. The suppression of oxidative damage and of the breakdown of lipid hydroperoxides in sickle cell membranes after incubation with desferrioxamine must arise by a chelation mechanism of the membrane-bound iron in some way, which would certainly imply that the iron is not in the state of irreversible haemichromes, contrary to the observations of Asakura et al. [81]. If this is the case and the state of the membrane-bound iron is chelatable, then in the unchelated form it may equally well be available for the Haber-Weiss reaction involving the iron-catalysed production of hydroxyl radicals from hydrogen peroxide and superoxide radicals, as suggested by Hebbel et al. [52], although we have no evidence here. Incorporation of antioxidant enzymes into the membrane system such as superoxide dismutase, catalase, and a combination of both, however, were ineffective in suppressing the incubation induced oxidative damage, whereas the chain-breaking antioxidant vitamin E [19, 20] decreased lipid peroxidative damage to the same extent as desferrioxamine after the 5-hour incubation.

These studies suggest the presence of lipid hydroperoxides in sickle cell membranes and the potential for the iron-catalysed production of alkoxy and peroxy radicals by the membrane-bound components of haemoglobin. These factors cause continued oxidative stress and amplify the membrane damage.

Iron, Iron Overload and Thalassaemia Major

Iron overload occurs whenever excess of the metal enters the body either as a result of repeated blood transfusion in patients with chronic refractory anaemias, such as thalassaemia major, or in increased gastrointestinal (GI) iron absorption as in idiopathic haemochromatosis [83, 84]. There is no specific mechanism for the excretion of excess iron from the body. Patients who are iron-loaded suffer from progressive hepatic, cardiac and endocrine dysfunction [85]. Iron accumulated via transfusion or the GI tract is transported via transferrin and stored bound to ferritin in most organs but predominantly in the liver, the reticuloendothelial system and the endocrine glands. Iron in this form is relatively inert and tissue damage only occurs when the capacity of the iron-binding proteins is exceeded, although it has been suggested that haemosiderin and iron-loaded transferrin may themselves be involved in iron-mediated damage in vivo [86] under certain conditions.

Of the 4 g of iron in the adult human, 2.5 g circulates in haemoglobin, 0.5 g is found in myoglobin, a small amount is in the various iron-containing proteins and the transport protein transferrin, the rest is found in the intracellular storage proteins ferritin and haemosiderin although there appear to be traces of iron also present in extracellular fluids [87]. There are four potential sources of iron in vivo:

(1) Iron taken up by the gut enters the plasma protein transferrin which binds iron (III) tightly on its two binding sites per molecule. Under normal conditions transferrin in the bloodstream is only 30% loaded with iron on average and so the amount of free iron salts available in the blood plasma would normally be zero.

(2) Any iron not required by cells is stored as ferritin (4,500 mol Fe/mol protein). Ferritin is found mainly in the liver, spleen and bone marrow but also to some extent in other tissues. Some ferritin is localised also in the blood plasma (normal serum ferritin levels 15–300 μg/l). The function of ferritin is to prevent excessive intracellular accumulation of non-protein-bound iron. In iron overload, synthesis of apoferritin is stimulated, therefore serum ferritin levels are a reflection of iron overload [88]. Iron enters ferritin as iron (III) which becomes oxidised by the protein to iron (III). Similarly, iron can be released from ferritin as iron (II) by the action of a number of biological reductants, e.g. ascorbate, superoxide [89].

(3) The intracellular mobile iron pool [90] contains the iron awaiting its role in the synthesis of the iron proteins haemoglobin, myoglobin, iron-sulphur proteins, etc. (although this is not relevant to the normal erythrocyte).

(4) The major location of iron in the body is in the blood in the form of the transport protein haemoglobin. In its oxygenated form, haemoglobin is not involved in the catalysis of damaging radical reactions. Oxyhaemoglobin, however, is really a methaemoglobin-superoxide radical complex (as has already been described) and disturbance of the oxygenation-deoxygenation cycle leading to the formation of methaemoglobin and superoxide can be the prime process leading to cell damage in many pathological situations.

Haem compounds are among the best catalysts for the breakdown of lipid hydroperoxides [40]. Furthermore, recent studies have also shown that exposure of haemoglobin in haemolysates to lipid peroxides or hydrogen peroxide can cause release of iron from it [91]. This released iron can catalyse lipid peroxidation and hydroxyl radical formation not only intra-erythrocytically but also in, for example, the inflamed rheumatoid joint, since haemoglobin from microbleeding can give up iron as it is attacked by hydrogen peroxide produced by activated neutrophils.

The thalassaemias are a group of inherited disorders of haemoglobin synthesis in which the production of a single type of globin gene is impaired. In β-thalassaemia major, the form most commonly found in London, the suppressed formation of β-globin chains, while α-chain synthesis proceeds normally, leads to an accumulation of unpaired α-chains. These precipitate and form intracellular inclusions which can be observed even in young nucleated erythroid cells [92]. Such inclusion bodies may interact with the membrane and form localised areas of membrane rigidity contributing towards a decrease in deformability of these thalassaemic cells [see refs. 92, 93, 56, 57 for reviews] and leading to their premature destruction in the spleen. This ineffective erythropoiesis causes the bone marrow to overproduce red cells and it expands to occupy all the space available to it, producing the enlargement of the skull characteristic of this disease.

In haemolytic anaemias such as sickle cell disease or haemoglobin H disease where the haemoglobin does not extensively precipitate, the iron from erythrocyte destruction is available for other developing red cells. In thalassaemia major, the formation of Heinz bodies exerts considerable oxidative stress as a result of increased production of oxygen radicals [36]. Isolated globin chains autoxidise much faster than the natural tetramer [36] producing breakdown products of oxidised haemoglobin chains and superoxide radicals. The iron from the α-chain inclusions digested within the reticuloendothelial cells seems to be moved preferentially into iron stores. This abnormal handling of iron leads to excessive GI iron absorption that can pose a serious clinical problem in the thalassaemia syndromes.

It is important to emphasise that the major cause of the cellular pathology in thalassaemia major is the α-chain imbalance. Chain imbalance is actually comparable in haemoglobin H disease and in thalassaemia major, yet the degree of ineffective erythropoiesis is much less in the former disease because only a small proportion of unpaired chains precipitate in the marrow–instead they associate to form soluble though non-functional β_4(haemoglobin H). These differences lead to haemolytic anaemia in haemoglobin H disease rather than the ineffective erythropoiesis in thalassaemia major. In β-thalassaemia it is the timing of precipitation and the nature of the precipitates which determine the timing of cell injury and death [94].

Untreated patients with thalassaemia major die in infancy unless regularly transfused. Regular blood transfusion causes iron loading leading to the appearance of low molecular weight iron complexes in the blood serum [95] and iron deposition mainly in the liver and spleen. Apoferritin synthesis is greatly enhanced in thalassaemic reticulocytes and relatively large amounts of iron in the form of the storage proteins, ferritin and haemosiderin, are present in the erythropoietic cells obtained from patients with this disease [96, 97].

The ways in which iron may cause tissue and cellular damage in such pathological conditions as thalassaemia are complex but in the blood such toxicity may arise from several sources: the excess availability of iron when plasma transferrin becomes saturated [98], from the accumulation of breakdown products of oxidised haemoglobin in the form of haemichromes close to the membrane surface of the thalassaemic erythrocyte [99] or from haem released from its natural globin anchor [100] and from the presence of ferritin-like iron residual from the thalassaemic reticulocyte [101]. Several inter-related factors, therefore, contribute towards the altered rheological behaviour and accelerated destruction of the thalassaemic erythrocyte. Biochemical abnormalities of these cells include: elevated levels of intracellular calcium [99] (especially in thalassaemia intermedia), a lowered level of potassium ions with a consequent loss of water and reduction in cell volume and osmotic fragility [102]. The final outcome is the shrinkage of the cell which would contribute towards the increased rigidity of the membrane [103a]. When individual Heinz bodies are present in cells they can be selectively removed during their passage through the spleen by physical pitting or plucking of the inclusion as it passes through the splenic cords. The resulting loss of surrounding membrane material which occurs as the Heinz bodies are removed leads to decreased surface area to volume ratio and therefore contributes further to reduced deformability and decreased cellular survival. The thalassaemic dacryocyte is shown in figure 6 [103b].

The formation of Heinz bodies exerts considerable oxidative stress as a result of the increased production of oxygen radicals [104]. The increased

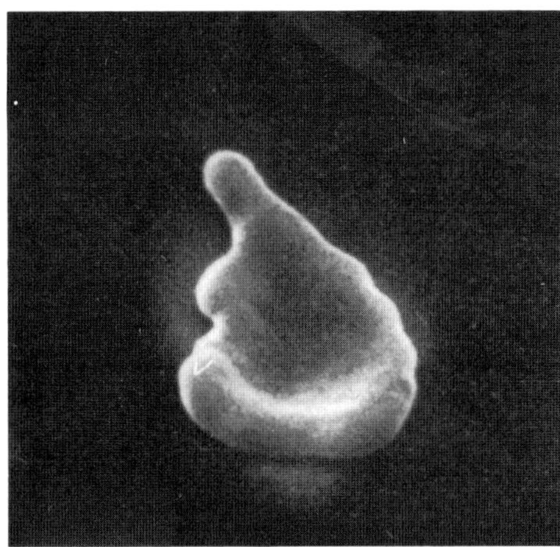

Fig. 6. The thalassaemic dacryocyte [103b].

susceptibility of the thalassaemic erythrocyte to oxidative stress is well-documented [105–107, 51]. Reduced phosphatidyl ethanolamine content in the membrane lipids, decreased arachidonic acid and increased palmitic acid [106], decreased lipid fluidity and lowered levels of free thiol groups in the membrane proteins [103a] all implicate peroxidative damage. Free radical-induced peroxidation of unsaturated fatty acids in the membrane phospholipids leads to the formation of aldehydic secondary lipid break-down products, some of which, e.g. malonaldehyde, may become involved in the polymerisation of the membrane proteins with important conse-quences for the mechanical properties of the cell and its survival time [108]; and e.g. 4-hydroxynonenal which has been implicated in the haemolysis of thalassaemic erythrocytes [107]. It has also been demonstrated that the changes in the quantity and distribution of sialic acids on the thalassaemic red cell surface causes increased phagocytosis in patients with thalassaemia intermedia [99]. How these relate to the biochemical, structural and func-tional alterations described is yet to be clarified but it has been shown [99] that cytoskeletons prepared from erythrocytes from splenectomised patients with thalassaemia intermedia contain more globin than normals.

In thalassaemia major and intermedia, the major threat to the erythro-cyte from iron-mediated oxidative damage seems to be from the iron in the intracellular inclusions in the thalassaemic erythrocyte, and from the possible presence of iron complexes which may be residual from the thalassaemic reticulocyte. It is possible that the increased production of the active oxygen species from isolated haemoglobin chains may lead to the accumulation of levels of peroxide which can interact with the haemoglobin producing free iron; alternatively or additionally active oxygen species can

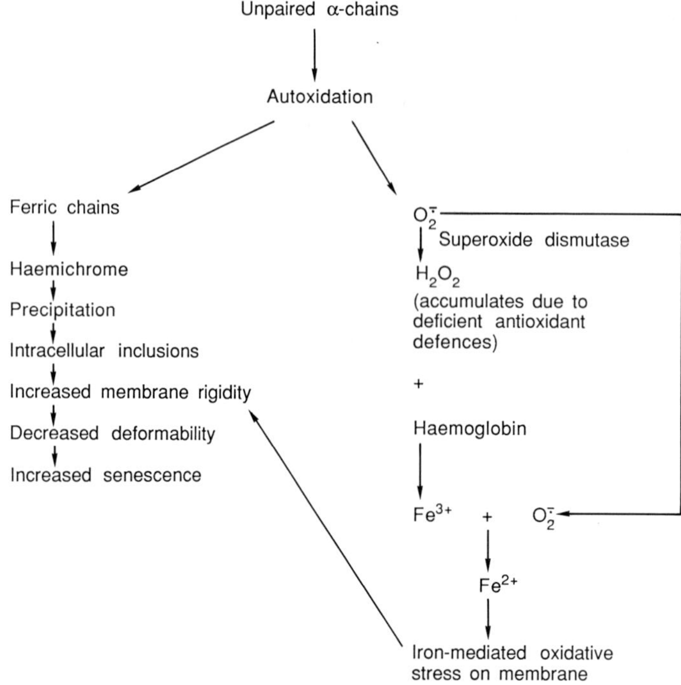

Fig. 7. A hypothetical scheme for the consequences of the accumulation of unpaired α-chains for the thalassaemic erythrocyte and its membrane.

interact with residual ferritin from the thalassaemic reticulocyte again releasing free iron. This iron would then be available to catalyse oxidative damage both intracellularly and in the membrane according to the scheme in figure 7. Modifications to the cytoskeletal assembly as well as to the lipid structure and organisation would result and, via modifications to the interactions with the intramembranous proteins, may alter the distribution of charge at the membrane surface.

Glucose-6-Phosphate Dehydrogenase Deficiency: Cellular and Membrane Modifications

The biochemical basis of the onset of the haemolytic crisis in G6PD-deficient subjects is still unknown. It has been established that a wide variety of oxidant drugs such as phenylhydrazine, primaquine, acetylsalicylic acid and the active consituents of the fava bean are capable of producing very reactive oxygen derivatives on autoxidation or via a linked oxidation between oxyhaemoglobin and the drug or its metabolites. These reactive species have the capacity to initiate lipid peroxidation leading to progressive membrane damage.

Oxidation of reduced glutathione accompanied by the action of glutathione peroxidase [109, 110] is the major mechanism by which hydrogen peroxide is metabolised in intact erythrocytes. Continuous reduction of the oxidised glutathione is dependent on glutathione reductase, which itself is $(NADPH + H^+)$-dependent. Since the reduced form of this enzyme is in short supply in individuals deficient in G6PD the resulting failure to maintain normal concentrations of reduced glutathione and the accompanying intracellular accumulation of hydrogen peroxide and oxidising radicals leads to the oxidation of free sulphydryl groups.

Fava beans contain a number of redox compounds that are capable of producing active oxygen intermediates on autoxidation [111, 112]. The active pyrimidine aglycones, divicine and isouramil, can interact with oxygen to produced hydrogen peroxide and this reaction can be maintained as a redox cycle in the presence of reduced glutathione which regenerates the reduced autoxidisable pyrimidine:

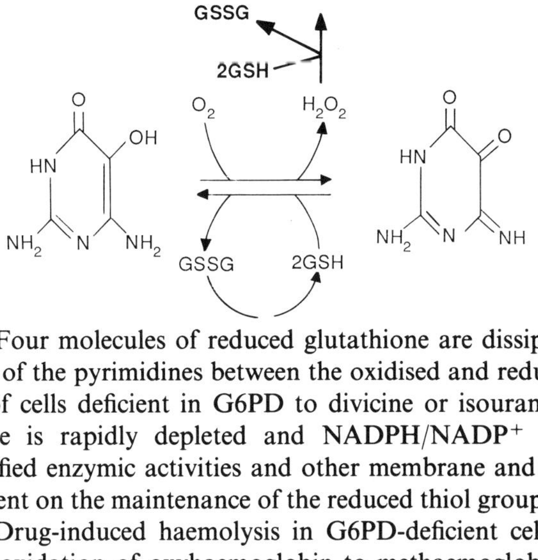

Four molecules of reduced glutathione are dissipated in one complete cycle of the pyrimidines between the oxidised and reduced states. On exposure of cells deficient in G6PD to divicine or isouramil the reduced glutathione is rapidly depleted and $NADPH/NADP^+$ ratios drop, causing modified enzymic activities and other membrane and cellular functions dependent on the maintenance of the reduced thiol groups will also be affected.

Drug-induced haemolysis in G6PD-deficient cells is often associated with oxidation of oxyhaemoglobin to methaemoglobin and the formation of particles of denatured haemoglobin and Heinz bodies. It is not clear whether the formation of methaemoglobin represents an important first step in Heinz body accumulation or whether this oxidation is merely coincidental. It is more certain that uncompensated oxidation of glutathione is central to the series of events that lead to the ultimate sequestration of affected cells by the reticuloendothelial system. During the haemolytic crisis a number of morphological and rheological modifications occur in the erythrocytes and biochemical changes in the membranes which contribute towards the decreased cell survival.

In this defect of glutathione metabolism the action of redox drugs (RH_2) is related to the ability of haemoglobin to activate oxygen to form ferric superoxide and function as an oxidase:

$$HbFe^{3+} - O_2^{\cdot -} + RH_2 + H^+ \rightarrow HbFe^{3+} + H_2O_2 + RH^{\cdot}$$

thus forming two potentially toxic species, hydrogen peroxide and the highly reactive drug radical as well as methaemoglobin, which becomes progressively oxidised and denatured to form Heinz bodies thereby contributing to the membrane damage, which could eventually lead to cell lysis.

The major questions of interest here are: What is the nature of the membrane damage in erythrocytes deficient in G6PD and what are the relative contributions from the oxygen radical-mediated processes and the membrane-bound haemoglobin towards the mechanism of cell lysis?

During the haemoltyic crisis disordered calcium homeostasis has been observed [113, 114] in G6PD-deficient erythrocytes. Significantly, impaired calcium-ATPase activity and a parallel marked increase of intracellular calcium levels is accompanied by a concomitant reduction in intracellular potassium levels. As described earlier, an elevated calcium concentration within the erythrocyte is expected to switch on a number of otherwise latent enzymic activities [115] which have a requirement for calcium ions such as (1) cytosolic proteinases [116], (2) transglutaminases, which have the potential for catalysing the formation of membrane protein aggregates (see below) thus causing accelerated red cell destruction [117], and (3) phospholipase C [118] which causes the enhanced conversion of phospho-inositides to diacylglycerol which could perturb the membrane architecture thus influencing other membrane functions. The simultaneous action of these enzymes is likely to alter both the shape and metabolic functions of the erythrocytes and consequently will influence their survival. Although these abnormalities undoubtedly contribute towards the haemolytic crisis, it is unlikely that they are the primary cause of the haemolysis leading to red cell destruction. (In vitro incubation [113] of normal and G6PD-deficient erythrocytes with divicine reproduced these events without haemolysis.

Apparent indicators of oxidative damage are the aggregates of high molecular weight proteins reported to have been observed in the membranes of G6PD-deficient erythrocytes from patients with chronic haemolysis [119], from patients with Mediterranean and A-variants associated with intermittent haemolysis [120] usually drug-induced, and from a patient with G6PD-Long Prairie [121]. In most cases, the majority of the aggregate is dissociable by thiol reducing agents showing that the major contributor is oxidant-induced intermolecular disulphide bond formation between reduced thiol groups [114,119]. The aggregates contain spectrin but no globin [119]. A small proportion of polypeptide aggregates are cross-linked via Ca^{2+}-activated transglutaminase especially in the older fraction of density-separated cells. A significant decrease in protein 3 is also observed in the older fraction of such cells from certain patients [114]. This latter component does not reappear after reduction suggesting that the potential effects of calcium-activated proteinases come into play here. Correlations have been shown between the formation of high molecular weight protein aggregates and decreased erythrocyte survival in model systems involving normal human erythrocytes both induced by sulphydryl-specific oxidants [122] and by oxygen radicals [123].

Other indicators of oxidant damage have been suggested by studies on the membrane lipids of erythrocytes from G6PD-deficient patients. Increased lipid peroxidation in the form of elevated thiobarbituric acid-reactive products have been observed in such membranes [120]. The consequences of lipid peroxidation and its implications in the membrane alterations associated with aging red cells by initiating changes in their mechanical properties is well-documented [5, 123, 124]. Furthermore, it has been demonstrated that the treatment of red cells with potential haemolytic agents results in a decrease in membrane lipid fluidity as well as in changes in the membrane surface charge accompanied by a decrease in cellular deformability. The drug-induced increased susceptibility to peroxidative damage of the membrane lipids susceptibility to peroxidative damage of the membrane lipids would lead to decreased fluidity, and studies in other systems have reported a correlation between modified membrane surface charge and oxidation status of the haemoglobin and of the membrane [125].

During acute haemolysis associated with G6PD-deficiency, Heinz bodies are characteristically formed [126]. There is reason to doubt, however, that this is the critical event leading to haemolysis since, in some patients with chronic haemolysis and no splenectomy, no evidence of Heinz body formation was obtained. Recent work has shown that in some patients during the favic crisis up to 50% of the red blood cell population constitute membrane cross-bonded red cells which are rapidly cleared from the circulation [127]. These observations have shown that during haemolysis the haemoglobin tends to be confined to one part of the cell leaving the other part as a transparent haemoglobin-free membrane. In this part of the cell the membrane appears tightly bonded as demonstrated by the observation that swelling did not peel apart the bonded membrane areas. Membrane modifications followed by squeezing in the microcirculation are proposed as being the cause of membrane cross-bonded red cells during the favic crisis. Membrane cross-bonding reduces the effective surface area of the erythrocytes causing loss of deformability and removal from the circulation.

Studies on the levels of antioxidant enzymes in erythrocytes from favic individuals [128] have shown increased superoxide dismutase levels and decreased glutathione peroxidase levels compared with normal individuals and non-favic G6PD-deficients. In conditions of decreased glutathione metabolism, excess of superoxide dismutation over hydrogen peroxide-detoxification enzymic activities may lead to a risk of subsequent interactions to produce hydroxyl radicals, under the right conditions, if the absolute amount of superoxide production is much higher than that in normal conditions. Similar results have been obtained on treatment of normal erythrocytes with high concentrations of divicine and ascorbate as a model for favism [128]. This treatment produces no haemolysis unless the treated cells are suspended in autologous plasma. These studies suggest a possible role for activated oxygen species in favism.

It has been proposed that G6PD-deficient erythrocytes have an impaired shape recovery mechanism which is a distinct property of the metabolic defect [129]. As morphologically abnormal erythrocytes are haemolysed more readily than normals [130], this would be expected to contribute towards the increased haemolysis in these defective cells particularly when under oxidative stress.

It is clear that drug-induced anaemia in G6PD-deficiency is in reality a disorder which reflects the impaired ability of cells to adequately detoxify hydrogen peroxide whatever its source [131]. Superoxide anions may be formed initially in many of the reactions which involve hydrogen peroxide accumulation. Utimately, erythrocytes are damaged as a consequence of the failure to maintain adequate levels of reduced glutathione for peroxide detoxification. While hydrogen peroxide itself may cause inhibition of enzymes, it is more probable that its cytotoxic effects involve its breakdown to yield hydroxyl radicals. It is to be presumed, as no clear evidence is available at present, that hydroxyl radical formation from peroxide is catalysed by iron from haemoglobin (possibly due to the influence of hydrogen peroxide itself on the haemoglobin [91]). Intravascular haemolysis is not easily observed. Hence hydroxyl radicals are proposed to exert their primary effects on the membrane components resulting in changes in their mechanical properties and rheological behaviour. Eventual sequestration of altered cells will result in anaemia without intravascular haemolysis.

References

1   Cohen G, Hochstein P: Generation of hydrogen peroxide in erythrocytes by haemolytic agents. Biochemistry 1964; 3: 895–900.

2   Kahane I, Schifter A, Rachmilewitz EA: Crosslinking of red blood cell membrane proteins induced by oxidative stress in thalassaemia major. FEBS Lett 1978;85:267–290.

3   Jacob HS, Winterhalter KH: The role of haemoglobin haem loss in Heinz body formation: studies with a partially haem-deficient and with genetically unstable haemoglobins. J Clin Invest 1970;49:2008–2016.

4   Jain SK, Hochstein: Polymerisation of membrane components in aging red blood cells. Biochem Biophys Res Commun 92:1980;247–252.

5   Hochstein P, Rice-Evans C: Lipid peroxide and membrane alterations in erythrocyte survival; in Yagi K (ed): Lipid Peroxides in Biology and Medicine. New York, Academic Press, 1982, pp 81–88.

6   Nagel RL, Ranney HM: Drug-induced denaturation of haemoglobin. Semin Haematol 1973;10:269–278.

7   Misra HP, Fridovich I: The generation of superoxide radical during the autoxidation of haemoglobin. J Biol Chem 1972;247:6960–6962.

8   Carrell RW, Winterbourh CC, Rachmilewitz EA: Activated oxygen and haemolysis. Br J Haematol 1975;30:259–264.

9   Eder HA, Finch C, McKee RW: Congenital methaemoglobinaemia. A clinical and biochemical study of a case. J Clin Invest 1940;28:265–272.

10 White JM: Haemoglobin variation; in Brock DJH, Mayo O (eds): Biochemical Genetics of Man. New York, Academic Press, 1972, pp 562–631.

11 Mohandas N, Chassis JA, Shohet SB: The influence of the membrane skeleton on red cell deformability, membrane material properties and shape. Semin Haematol 1983;20:225–242.

12 Lux S, Shohet SB: The erythrocyte membrane: biochemistry. Hosp Pract 1984; 19(10):77–83.

13 Lacelle PL: Alteration in membrane deformability in haemolytic anaemias. Semin Haematol 1970;7:355–371.

14 Mohandas N, Phillipis WH, Bessis M: Red blood cell deformability and haemolytic anaemias. Semin Haematol 1970;16:95–114.

15 Marchesi V: The cytoskeletal system of red blood cells. Hosp Pract 1985;20(11):113–131.

16 Fridovich I: Oxygen radicals, hydrogen peroxide and oxygen toxicity; in Pryor WA (ed): Free Radicals in Biology. New York, Academic Press, 1976, Vol I, pp 239–277.

17 Fridovich I: Superoxide radical: an endogenous toxicant. Annu Rev Pharmacol Toxicol 1983;23:239–257.

18 Cohen G, Hochstein P: Glutathione peroxidase: The primary agent for the elimination of hydrogen peroxide in erythrocytes. Biochemistry 1963;2:1420–1428.

19 Tappel AL: Vitamin E as the biological lipid antioxidant. Vitam Horm 1962;20:483–510.

20 Diplock AT: The role of vitamin E in biological membranes; in Biology of Vitamin E. CIBA Fdn Symp No 101. Amsterdam, Elsevier, 1983; pp 45–53.

21 Osaki S, Johnson DA, Frieden E: Possible significance of ferrous oxidase activity of caeruloplasmin in normal human serum. J Biol Chem 1966;241:2746–2751.

22 Boyer RF, Schori BE: the Incorporation of iron into apoferritin as mediated by caeruloplasmin. Biochem Biophys Res Commun 1983;116:244–250.

23 Winterbourn CC: Comparison of superoxide with other reducing agents in the biological production of hydroxyl radicals. Biochem J 1979;182:625–628.

24 Winterbourn CC: Hydroxyl radical production in body fluids. Roles of metal ions, ascorbate and superoxide. Biochem J 1981;198:125–131.

25 Rowley DA, Halliwell B: Formation of hydroxyl radicals from hydrogen peroxide and iron salts by superoxide- and ascorbate-dependent mechanisms: relevance to the pathology of rheumatoid disease. Clin Sci 1983;64:649–653.

26 Rowley DA, Halliwell B: Superoxide-dependent formation of hydroxyl radicals in the presence of thiol compounds. FEBS Lett 1982;138:33–36.

27 Searle A, Tomasi A: Hydroxyl free radical production in iron-cysteine solutions and protection by zinc. J Inorg Chem 1982;17:161–166.

28 Rowley DA, Halliwell B: Superoxide-dependent formation of hydroxyl radicals from NADH and NADPH in the presence of iron salts. FEBS Lett 1982;142:39–41.

29 Fenton HJH: Oxidation of tartatic acid in the presence of iron, J Chem Soc 1894;23:899–910.

30 Haber F, Weiss J: The catalytic decomposition of hydrogen peroxide by iron salts. Proc Soc Lond Series A 1934;147:332–351.

31 Halliwell B: Free radicals, oxygen toxicity and aging; in Sohal RS(ed): Age Pigments. Amsterdam, Elsevier, 1981, pp 1–62.

32 Slater TF: Free radical mechanisms in tissue injury. Biochem J 1984;222:1–15.

33 Gutteridge JMC, Rowley DA, Halliwell B: Superoxide-dependent formation of hydroxyl radicals in the presence of iron salts. Biochem J 1981;199:263–265.

34 Peisach J, Blumberg WE, Rachmilewitz EA: The demonstration of ferrihaemochrome intermediates in Heinz body formation following the reduction of oxyhaemoglobin A by acetylphenylhydrazine. Biochim Biophys Acta 1975;393:404–418.

35 Heniz R: Über Blut Degeneration und Regeneration. Beitr Pathol 1901;29:299.

36 Brunori M, Falcioni G, Fioreti E, et al: Formation of superoxide in the autoxidation of isolated chains of haemoglobin and its involvement in haemichrome precipitation. Eur J Biochem 1975;53:99–104.

37 Girotti A: Mechanisms of lipid peroxidation. J Free Rad Biol Med 1985;1:87–95. (1985).

38 Halliwell B, Gutteridge JMC: The importance of free radicals and catalytic metal ions in human disease. Mol Aspects Med 1985;8:89–193.

39 Kappus H: Lipid peroxidation: mechanisms, analysis, enzymology and biological relevance; in Sies H (ed): Oxidative Stress. New York, Academic Press, 1985, pp 273–310.

40 O'Brien P: Intracellular mechanisms for the decomposition of a lipid peroxide by metal ions, haem compounds and nucleophiles. Can J Biochem 1969;47:485–492.

41 Halliwell B, Gutteridge JMC: Oxygen toxicity, oxygen radicals, transition metals and disease. Biochem J 1984;219:1–14.

42 Tappel AL: Measurement of and protection from in vivo lipid peroxidation; in Pryor WA (ed): Free Radicals in Biology. Academic Press, 1980. vol IV, pp 1–45.

43 Esterbauer H: Aldehydic products of lipid peroxidation; in McBrien DCH, Slater TF (eds): Free Radicals, Lipid Peroxidation and Cancer. New York, Academic Press, 1982, pp 101–128.

44 Esterbauer H: Lipid peroxidation products: formation, chemical properties and biological activities; in Poli G, Cheeseman KH, Dianzani MU, Slater TF (eds): Free Radicals and Liver Injury. IRL Press, 1985, pp 29–47.

45 Esterbauer H, Cheeseman KH, Dianzani MU, et al: Separation and characterisation of the aldehydic products of lipid peroxidation stimulated by ADP-Fe^{2+} in liver microsomes. Biochem J 1982;208:129–140.

46 Kappus H, Sies H: Toxic drug effects associated with oxygen metabolism: redox cycling and lipid peroxidation. Experientia 1981;37:1233–1241.

47   Tappel AL, Dillard CJ: In vivo lipid peroxidation measurement by exhaled pentane and protection by vitamin E. Fed Proc 1981;40:174–178.
48   Wolff SP, Dean RT: Fragmentation of proteins by free radicals and its effect on their susceptibility to exzymic hydrolysis. Biochem J 1986;234:399–403.
49   Wolff SP, Garner A, Dean RT: Free radicals, lipids and protein degradation. Trends Biochem Sci 1986;11:27–31.
50   Rice-Evans C, Omorphos S, Baysal E: Sickle cell membranes and oxidative damage. Biochem J 1986;237:265–269.
51   Chiu D, Lubin B, Shohet SB: Peroxidative reactions in red cell biology; in Pryor WA (ed): Free Radicals in Biology. New York, Academic Press, 1982, vol V, pp 115–160.
52   Hebbel RP, Eaton JW, Balasingham M, et al: Spontaneous oxygen radical generation by sickle erythrocytes. J Clin Invest 1982;70:1253–1259.
53   Dean J, Schechter AN: Sickle cell anaemia: molecular and cellular bases of therapeutic approaches. N Engl J Med 1978;299:804–811.
54   Lux SE, John KM, Karnovsky MJ: Irreversible deformation of the spectrin-actin lattice in irreversibly sickled cells. J Clin Invest 1976; 59:955–963.
55   Lux SE, John KM: The role of spectrin and actin in irreversibly sickled cells: Unsickling of irreversibly sickled ghosts by conditions which interfere with spectrin-actin polymerisation; in Caughey W (ed): Biochemical and Clinical Aspects of Haemoglobin Abnormalities. New York, Academic Press, 1978, pp 335–353.
56   Palek J: Red cells membrane injury in sickle cell anaemia. Br J Haematol 1977;35:1–9.
57   Rice-Evans C, Dunn MJ: Erythrocyte deformability and disease. Trends Biochem Sci 1982;7:282–286.
58   Eaton JW, Skelton TO, Swofford HS, et al: Elevated erythrocyte calcium in sickle cell disease. Nature 1973;246:105–106.
59   Glader BE, Lux SE, Muller-Soyano A, Energy reserve and cation composition in irreversibly sickled cells in vivo. Br J Haematol 1978;40:527 532.
60   Chiu D, Lubin B, Shohet SB: Erythrocyte membrane reorganisation during the sickling process. Br J Haematol 1979;41:223–234.
61   Rice-Evans C, Bruckdorfer KR, Dootson G: Studies on the altered membrane characteristics of sickle cells. FEBS Lett 1978;94:81–86.
62   Hebbel RP, Schwartz RS, Mohandas N: The adhesive sickle erythrocyte: causes and consequences of abnormal interactions with endothelium, monocytes, macrophages and model membranes. Clin Haematol 1985;14:141–153.
63   Siegel DL, Goodman SR, Branton D: The effect of endogenous proteases on the spectrin binding proteins of human erythrocytes. Biochim Biophys Acta 1980;598:517–527.
64   Allan D, Michell RH: Calcium ion-dependent diacyl glycerol accumulation in erythrocytes is associated with microvesiculation but not with efflux of potassium ions. Biochem J 1977;166:495–499.
65   Bookchin RM, Ortiz OE, Lew VL: Red cell calcium transport and mechanisms of dehydration in sickle cell anaemia; in Beuzard Y, Charache S, Galacteros F (eds): Approaches to the Therapy of Sickle Cell Anaemia. Paris, INSERM, 1986 pp 291–299.
66   Allan D, Limbrick AR, Thomas P, et al: Microvesicles from sickle erythrocytes and their relation to irreversible sickling. Br J Haematol 1981;47:383–390.
67   Allan D, Limbrick AR, Thomas P, et al: Release of spectrin-free spicules on reoxygenation of sickle erythrocytes. Nature 1982;295:612–613.
68   Westerman M, Cole ER, Wu K: The effect of spicules obtained from sickle red cells on clotting activity. Br J Haematol 1984;56:557–562.
69   Wagner GM, Schwartz RS, Chiu D, et al: Membrane phospholipid organisation and vesiculation of erythrocytes in sickle cell anaemia. Clin Haematol 1985;14:183–200.

70 Franck PFH, Bevers EM, Lubin B, et al: Uncoupling, of the membrane skeleton from the lipid bilayer. The cause of accelerated phospholipid flip-flop leading to an enhanced procoagulant activitiy of sickled cells. J Clin Invest 1985;75:183–190.

71 Hebbel RP, Miller WJ: Phagocytosis of sickle erythrocytes: immunologic and oxidative defects of haemolytic anaemia. Blood 1984;64:733–741.

72 Rice-Evans C, Omorphos S: Hydrazine and sickle cells. Biochem Soc Trans 1983;11:180–181.

73 Das SK, Nair RC: Superoxide dismutase, glutathione peroxidase, catalase and lipid peroxidation of normal and sickled erythrocytes. Br J Haematol 1980;44:87–92.

74 Chiu D, Lubin B: Abnormal vitamin E and glutathione peroxidase levels in sickle cell anaemia. J Lab Clin Med 1979;94:542–548.

75 Rice-Evans C, Omorphos S, Baysal E: Sickle cell pathology: Is the membrane important? in Rice-Evans C (ed): Free Radicals, Cell Damage and Disease. London, Richelieu Press, 1986, pp 149–166.

76 Rice-Evans C, Omorphos SC: Free radical-induced damage in sickle erythrocytes; in Beuzard Y, Charache S, Galacteros T (eds): Approaches to the Therapy for Sickle Cell Anaemia. Paris, INSERM, 1986, pp 329–336.

77 Halliwell B: Superoxide-dependent formation of hydroxyl radicals in the presence of iron chelates. FEBS Lett 1978;92:321–326.

78 Graf E, Mahoney JR, Bryant RG, et al: Iron-catalysed hydroxyl radical formation. J Biol Chem 1984;259:3620–3624.

79 Shaklai N, Sharma VS, Ranney HN: Interaction of sickle cell haemoglobin with erythrocyte membranes. Proc Natl Acad Sci USA 1981;78:65–68.

80 Rice-Evans C, Omorphos S, Baysal E, et al: Free radicals, sickle cells and antioxidants; in Beuzard Y, Charache S, Galacteros T (eds): Approaches to the Therapy for Sickle Cell Anaemia. Paris, INSERM, 1986, pp 315–328.

81 Asakura T, Minakata M, Adachi K, et al: Denatured haemoglobin in sickle erythrocytes. J Clin Invest 1977;59:633–640.

82 Gutteridge JMC, Richmond R, Halliwell B: Inhibition of the iron-catalysed formation of hydroxyl radicals from superoxide and of lipid peroxidation by desferrioxamine. Biochem J 1979;184:469–472.

83 Modell B, Beck J: Long-term desferrioxamine therapy in thalassaemia. Ann NY Acad Sci 1974;232:201–210.

84 Jacobs A: Iron overload–clinical and pathological aspects. Semin Hematol 1977;14:89–113.

85 Ley TJ, Griffith P, Ninhuis AW: Transfusion siderosis and chelation therapy. Clin Haematol 1982;11:437–454.

86 O'Connell M, Halliwell B, Moorhouse CP, et al: Formation of hydroxyl radicals in the presence of ferritin and haemosiderin. Biochem J 1986;234:727–731.

87 Aust SD, Morehouse LA, Thomas CE: Role of metals in oxygen radical reactions. J Free Rad Biol Med 1985;1:3–25.

88 Skull GE, Theil EC: Regulation of ferritin mRNA: a possible gene-sparing phenomenon. J Biol Chem 1983;258:7921–7923.

89 Thomas CE, Morehouse LA, Aust SD: Ferritin and superoxide-dependent lipid peroxidation. J Biol Chem 1985;260:3275–3280.

90 Octave J-N, Schneider Y-J, Trouet A, et al: Iron uptake and utilisation by mammalian cells. Trends Biochem Sci 1983;8:217–220.

91 Gutteridge JMC: Iron promoters of the Fenton reaction can be released from haemoglobin by peroxides. FEBS Lett 1986;201:291–295.

92 Polliack A, Rachmilewitz EA: Ultrastructural studies in thalassaemia major. Br J Haematol 1973;24:319–326.

93 Rice-Evans C, Chapman D: Red blood cell biomembrane structure and deformability. Scand J Clin Lab Invest 1981;156:99–110.

94 Modell B, Berdoukas V: The Clinical Approach to Thalassaemia. New York, Grune & Stratton, 1984.

95 Hershko C, Graham G, Bates GW, et al: Non-specific serum iron in thalassaemia: an abnormal serum iron fraction of potential toxicity. Br J Haematol 1978;40:255–263.

96 Eylar EH, Matioli G: Apoferritin synthesis in human erythroid cells in thalassaemia. Nature 1965;208:661–664.

97 Polliack A, Yataganas X, Thorell B, et al: An electron microscopic study of the nuclear abnormalities in erythrocblasts in thalassaemia major. Br J Haematol 1974;26:201–204.

98 Worwood MJ: Iron and haemochromatosis. J Inher Metab Dis 1983;6:63–69.

99 Rachmilewitz EA: Erythrocyte membrane alterations in thalassaemia. Clin Haematol 1985;14:163–182.

100 Shaklai N, Shviro Y, Rabizadeh E, et al: Accumulation and drainage of haemin in the red cell membrane. Biochim Biophys Acta 1985;821:355–356.

101 Bauminger ER, Cohen SG, Offer S, et al: Quantitative studies of ferritin-like iron in erythrocytes of thalassaemia, sickle cell anaemia and haemoglobin Hammersmith by Mossbauer spectroscopy. Proc Natl Acad Sci USA 1979;76:939–943.

102 Kahane I, Rachmilewitz EA: Alteration in the red blood cell membrane and the effect of vitamin E on osmotic fragility in thalassaemia major. Isr J Med Sci 1976;12:11–15.

103a Rice-Evans C, Johnson A, Flynn DM: Red cell membrane abnormalities in thalassaemia major. FEBS Lett 1980;119:53–57.

103b Bessis M: in Bessis M, Weed RI, Leblond P (eds): Red Cell Shape. Berlin, Springer, 1973.

104 Winterbourn CC, McGrath BM, Carrell RW: Reactions involving superoxide and normal and unstable haemoglobins. Biochem J 1976;155:493–502.

105 Stocks J, Offerman L, Modell CB, et al: The susceptibility to autoxidation of human red cell lipids in health and disease. Br J Haematol 1972;23:713–724.

106 Rachmilewitz EA, Lubin B, Shohet SB: Lipid membrane peroxidation in thalassaemia major. Blood 1976;47:494–505.

107 Poli G, Ramenghi U, David O, et al: Lipid peroxidation in red blood cells obtained from thalassaemic subjects and exposed to oxidative stress; in Rice-Evans C (ed): Free Radicals, Cell Damage and Disease. London, Richelieu Press, 1986, pp 187–199.

108 Hochstein P, Jain SK, Rice-Evans C: The physiological significance of oxidative perturbations in erythrocyte membrane lipids and proteins; in Brewer G (ed): The Red Cell. New York, Liss, 1981, pp 449–459.

109 Cohen G, Hochstein P: Glutathione peroxidase: the primary agent for the elimination of hydrogen peroxide from the erythrocyte. Biochemistry 1963;2:1420–1428.

110 Cohen G, Hochstein P: Glucose-6-phosphate dehydrogenase and the detoxification of hydrogen peroxide in human erythrocytes. Science 1961;134:1756–1757.

111 Mager J, Glaser G, Razin A, Metabolic effects of pyrimidine derived from fava bean glycosides on human erythrocytes deficient in glucose-6-phosphate dehydrogenase. Biochem Biophys Res Commun 1965;20:235–240.

112 Mager J, Chevion M, Glaser G: in Liener IE(ed): Toxic Constituents of Plant Food-stuffs. New York, Academic Press, 1980, pp 265–284.

113 De Flora A, Benatti U, Guida L, et al: Favism: disordered erythrocyte calcium homeostasis. Blood 1985;66:294–297.

114 Turrini F, Naitana A, Mannuzzu L, et al: Increased red cell calcium, decreased calcium ATPase and altered membrane proteins during fava bean haemolysis in glucose-6-phosphate dehydrogenase-deficient individuals.

115 Benatti U, Guida L, Forteloni G, et al: Impairment of the calcium pump of human erythrocytes by divicine. Arch Biochem Biophys 1985;239:334–341.

116 Waxman L: Methods Enzymol 1981;80:664–680.

117 Lorand L, Weissman LB, Epel DL, Role of intrinsic transglutaminases in the calcium-mediated cross-linking of proteins. Proc Natl Acad Sci USA 1976;73:4479–4481.

118 Allan D, Thomas P, Michell RH: Rapid transbilayer diffusion of 1,2-diacly glycerol and its relevance to the control of membrane curvature. Nature 1978;276:289–290.

119 Johnson GJ, Allen DW, Cadman S, et al: Red cell membrane polypeptide aggregates in G6PD mutants with chronic haemolytic disease. A clue to the mechanism of haemolysis. N Engl J Med 1979;301:522–527.

120 Rice-Evans C, Rush J, Omorphos SC, et al: Erythrocyte membrane abnormalities in glucose-6-phosphate dehydrogenase deficiency of the Mediterranean and A-types. FEBS Lett 1981;136:148–152.

121 Allen DW, Johnson GJ, Cadman S, et al: Membrane polypeptide aggregates in glucose-6-phosphate dehydrogenase-deficient and in vitro aged red blood cells. J Lab Clin Med 1978;91:321–327.

122 Maeda M, Kon K, Imaizumi K, et al: Alteration of the rheological properties of human erythrocytes by cross-linking membrane proteins. Biochim Biophys Acta 1983;735;104–112.

123 Corry WD, Meiselman H, Hochstein P: t-Butyl hydroperoxide-induced changes in the physicochemical properties of human erythrocytes. Biochim Biophys Acta 1980;597:224–234.

124 Pfeffer SR, Swislocki NI: Role of peroxidation in erythrocyte aging. Mech Ageing Dev 1982;18:355–367.

125 Rice-Evans C, Baysal E, Pashby DP, et al: t-Butyl hydroperoxide-induced perturbations of human erythrocytes as a model for oxidant stress. Biochim Biophys Acta 1985;815:426–432.

126 Beutler E: in Stanbury et al. (eds): The Metabolic Basis of Inherited Disease. New York, McGraw-Hill, 1983, pp 1629–1654.

127 Fischer TM, Meloni T, Pescarmona G, et al: Membrane crossbonding M in red cells in favic crisis: a missing link in the mechanism of extravascular haemolysis. Br J Haematol 1985;59:159–169.

128 Mavelli I, Ciriola MR, Rossi L, et al: Favism: a haemolytic disease associated with increased superoxide dismutase and decreased glutathione peroxidase activities in red blood cells. Eur J Biochem 1984;139:13–18.

129 Alhanaty E, Snyder M, Sheetz MP: G6PD-deficient erythrocytes have an impaired shape recovery mechanism. Blood 1984;63:1198–1202.

130 Cosgrove P, Sheetz MP: Effect of cell shape on extravascular haemolysis. Blood 1982;59:412–427.

131 Hochstein P, Rice-Evans C: Glucose-6-phosphate dehydrogenase deficiency: mechanisms of oxidant-induced haemolysis; in Rice-Evans C (ed): London, Richelieu Press, 1986, pp 167–174.

Catherine Rice-Evans, PhD, BSc, Department of Biochemistry and Chemistry, Royal Free Hospital School of Medicine, Rowland Hill Street, London, NW3 2PF (UK)

2 Neutrophil Oxidants: Production and Reactions

Catherine C. Winterbourn

Neutrophils are the phagocytic cells that provide the first line of defence against invading micro-organisms. They are also major contributors to the inflammatory response, collecting at sites of inflammation where they act in a phagocytic or secretory capacity. Neutrophils respond to inflammatory stimuli in a number of ways. They undergo chemotaxis, adhesion, release of cytoplasmic granules containing lytic and digestive enzymes, synthesis and release of leukotrienes, as well as production of the superoxide radical (O_2^-) and other reactive oxygen species. The purpose of this review is to consider how these reactive oxygen species are involved in the microbicidal and inflammatory activities of neutrophils. It covers the nature of the species formed and the mechanism of formation, how oxidant production relates to other cell responses, and the evidence for oxidant involvement in various cytotoxic or inflammatory reactions.

For more than 50 years it has been known that engulfment of particles, e.g. micro-organisms, by neutrophils is accompanied by a burst of O_2 consumption [1]. This O_2 uptake (respiratory burst) is not due to conventional respiration. Early studies showed that it is accompanied by stimulation of the hexose monophosphate shunt, and results in the production of H_2O_2 [2, 3]. It is now known, through studies employing superoxide dismutase, that the initial product of O_2 reduction is O_2^- [4–6]. The respiratory burst is essential for efficient microbicidal activity. Under anaerobic conditions, neutrophils phagocytose but do not kill bacteria efficiently [7], and in chronic granulomatous disease, a disease manifested by recurrent infections and an inability of the neutrophils to kill certain micro-organisms, the respiratory burst is completely absent [8, 9].

Whether products of the respiratory burst contribute positively to other aspects of the inflammatory response is less certain. They are generally regarded as undesirable in the context of chronic inflammation, with potential to inflict damage on surrounding cells and structural matrix constituents. However, they could play a positive role in processes such as tissue remodelling and removal of cell debris, or possibly in controlling the activity of the neutrophils themselves and the extent of the inflammatory response. While the picture is far from complete, current knowledge allows us to make some assessment of these roles.

Table 1. Mediators of neutrophil superoxide production [accumulated from ref. 10–21]

Opsonized particles (e.g. micro-organisms)	cis-Polyunsaturated fatty acids,
Immune complexes (Fc receptor)	e.g. arachidonate
Complement fragment C5a	Cerebroside sulphate
N-Formylated peptides (bacterial origin),	Fluoride
e.g. fMet-Leu-Phe	Leukotriene B_4
Ca-Ionophore A23187	Tumour necrosis factor
Lectins, e.g. concanavalin A	Platelet-activating factor
Phorbol esters, e.g. PMA	Platelet-derived growth factor
Diacylglycerol	

Neutrophil Stimulation

Normal circulating (resting) neutrophils maintain a low level of O_2 consumption and produce insignificant amounts of O_2^-. In response to various mediators, a complex set of events occur within the cells, collectively termed 'stimulus-response coupling' which switch on a number of processes including the enzymatic conversion of O_2 to O_2^- [10–12]. The list of stimuli is varied and ever-expanding (table 1). It includes not only particulate stimuli such as immunoglobulin-coated bacteria, but also soluble stimuli and inflammatory mediators such as immune complexes, C_{5a} and leukotriene B_4. The varied nature of these stimuli is a clear indication that the role of the neutrophil in the inflammatory response extends beyond that of a phagocyte.

The mechanism of stimulus response coupling in the neutrophil is still not completely understood, but some features are emerging. The sequence of events, starting with binding of the stimulus to a membrane receptor, followed by phosphatidyl inositol turnover, Ca^{2+} mobilization and protein kinase C activation, appears to follow the general pathway of receptor-mediated responses [22, 23], and is summarized in figure 1. A full description is beyond the scope of the current article, and for this, readers are referred elsewhere [10, 12, 24–27]. Most of the stimuli listed in table 1 bind to specific receptors on the neutrophil membrane. Receptors for the F_c region of immunoglobulin, C_{5a}, LTB_4 and fMet-Leu-Phe have been identified. The calcium ionophore A_{23187} can bypass the membrane response by directly allowing stimulation of Ca^{2+}-dependent steps. Phorbol esters, such as phorbol myristate acetate (PMA), are structural analogues of diacylglycerol and can activate protein kinase C directly.

There are several issues which are yet to be resolved. Firstly, there is more than one route to activation of the NADPH oxidase [24, 28–31].

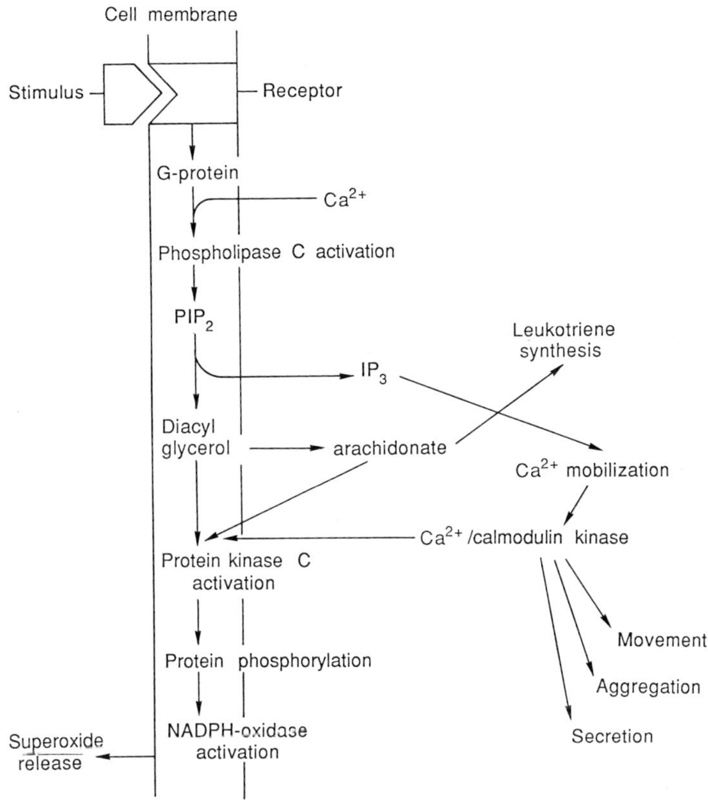

Fig. 1. Proposed scheme for the stimulus-response coupling mechanism of neutrophils [adapted from ref. 12, 22, 23]. G protein = GTP binding protein inhibited by pertussis toxin; PIP_2 = phosphatidylinositol 4,5-biphosphate; IP_3 = inositol 1,4,5-triphosphate.

Evidence, mainly from studies of the time course of activation with different stimulants, and also using specific inhibitors, imply the existence of Ca^{2+}-dependent and independent routes, the former possibly independent of protein kinase C, but the control and relative significance of the two routes are not yet known. Seconly, it is not understood how the neutrophil switches on some responses but not others. More is known about the respiratory burst than other responses, but as is shown in figure 1, all are linked by the same series of events. Yet stimulation can be specific. It is possible to activate the oxidase without degranulation, or with compounds such as fMet-Leu-Phe, to see only chemotaxis and adhesion at low concentrations, while higher concentrations stimulate a full set of responses [10, 11]. Furthermore, neutrophils contain at least three types of granule (azurophil, specific, C-particle), each under separate control [11]. Particulate or surface stimuli (e.g. surface-bound immune complexes) release all granule types, whereas soluble stimuli tend to cause very little azurophil degranulation [10, 11, 32, 33]. C-particles, on the other hand, are released at stimulant concentrations below the threshold required for oxidase activity or release of other granules [11]. Addition of cytochalasin B to neutrophils prior to stimulants such as fMet-Leu-Phe or immune complexes enhances degranulation and increases the detectable O_2^- production. Cytochalasin B interferes with microfilament formation and prevents complete closure of phagosomes, and this was originally proposed as the reason for its effects. Some effects of cytochalasin B cannot be explained solely on this basis, however, and it may also modify neutrophil responsiveness to some stimuli [11, 15, 16].

As distinct from acting as direct stimulants of neutrophil responses, a number of inflammatory mediators can 'activate' the cells so that their responses are augmented on subsequent stimulation. The term activation is used in the same context as that commonly applied to macrophages that have been immunologically primed either in vivo or in vitro (e.g. with lymphokines) to enhance their phagocytic and antimicrobial activity [34]. A similar mechanism appears to operate in neutrophils, even though they are much shorter lived and have limited ability to synthesize protein. Granulocyte-macrophage colony-stimulating factor [35, 36], interferon-γ [37], tumour necrosis factor [37, 38], platelet-activating factor [39], and endotoxin (bacterial lipopolysaccharide) [40–42] activate neutrophils. Enhanced adhesion, O_2^- production, degranulation and cytotoxicity have been measured. The mechanism of activation is not fully understood, but may involve an alteration in receptor numbers or affinity or enhanced activation of latent enzymes.

Similar mechanisms are likely to be operative in the phenomenon termed 'priming' [43–46]. Priming of neutrophils by exposure to low concentrations of fMet-Leu-Phe or complement peptides enhances their ability to generate O_2^- (and exhibit enhanced bactericidal activity) in response to a second stimulus. In this case, kinetic studies have shown that chemotactic peptides alter the V_{max} but not K_m of the oxidase [46], but the mechanism is again unclear. There is evidence that such priming occurs in vivo. Neutrophils isolated from patients with acute bacterial infections have increased O_2^--generating capacity [47]. Using a fluorescence method for measuring oxidant production and separating individual cells by flow cytometry, such patients show two populations of cells, one with normal and one with up to threefold greater generating activity [48].

Inhibition of Neutrophil Responses

Neutrophil stimulus responses are sensitive to a variety of inhibitors. Of particular interest is the inhibition by a wide range of both steroidal and non-steroidal anti-inflammatory drugs. Prevention of chemotaxis and adhesion, as well as O_2^- production, have been observed [49–57]. This raises the possibility that their effects on neutrophil function may contribute to the anti-inflammatory activity of these drugs, and also that they may be useful in the treatment of other conditions in which neutrophils are implicated [58].

How these inhibitors act is not known. Some are effective only with certain stimuli, and may compete for receptor binding sites [51, 52, 54, 56]. Others are more generally effective, and some that inhibit responses to fMet-Leu-Phe also do so with immune complexes, but only at a much higher drug concentration [52, 53]. The nature of the neutrophil stimulus, and the availability of appropriate concentrations of inhibitor, therefore, would appear to be key determinants of the physiological relevance of this mode of action of the drugs.

Superoxide Generation – The NADPH-Oxidase

The respiratory burst enzyme has received considerable attention since its discovery, and has always been controversial. For many years it was debated whether NADH or NADPH was the source of reducing equivalents for O_2 reduction. This issue is now resolved in favour of NADPH [12, 59, 60] converting O_2 to O_2^- according to the reaction

$$NADPH + 2O_2 \rightarrow NADP^+ + H^+ + 2O_2^-$$

The reaction takes place at the cell surface, with the NADPH being consumed intracellularly, but the O_2^- released to the outside [reviewed in refs. 12, 60, 61]. This includes release into phagosomes or vacuoles, since these constitute invaginations of the plasma membrane, and the inner surface of the phagosome is equivalent to the outer surface of the cell. The NADPH is regenerated from $NADP^+$ via the hexose monophosphate shunt, which accounts in part for the increased activity of this pathway when the cells are stimulated. It is also stimulated because some of the H_2O_2, produced via dismutation of O_2^-, enters the cell and reacts with reduced glutathione and glutathione peroxidase. Reduction of the oxidized glutathione (GSSG) formed requires a further input of NADPH.

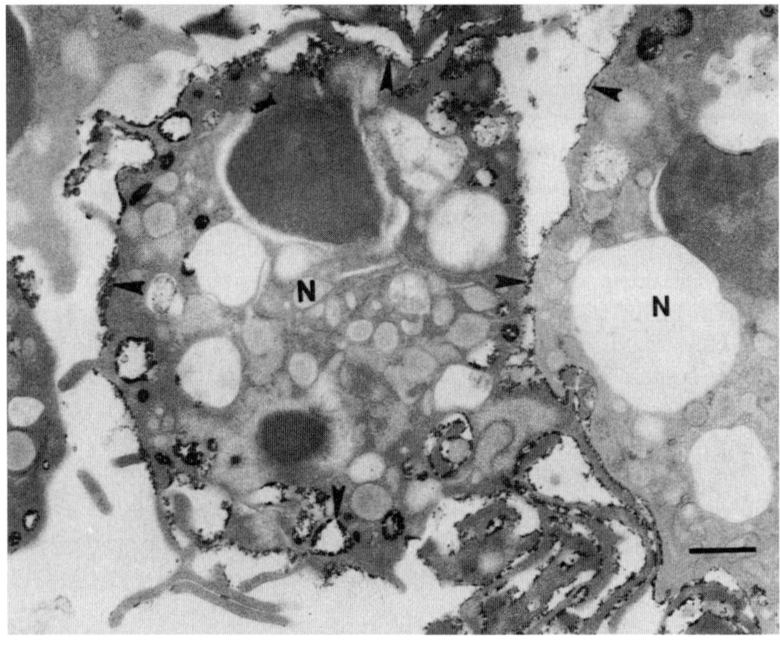

Fig. 2. Electron micrographs showing the site of H_2O_2 production. **a** Neutrophils stimulated with PMA. **b** Neutrophil stimulated by IgG aggregates within glomerular basement membrane. H_2O_2 was localized with cerium chloride by the method described in ref. [63]. Results are from ref. [64] (reproduced with permission from J Clin Invest). N = Neutrophil; GBM = glomerular basement membrane; arrow heads indicate cerium peroxide deposit.

The constitution of the oxidase system is still unresolved. A detailed discussion of this topic is beyond the scope of this review, but has received recent coverage by other authors [12, 24, 62–64]. It is generally agreed that the NADPH oxidase is an electron transport chain which assembles in the plasma membrane following stimulation of the cells. Although it has proved extremely difficult to purify in active form, two components are reasonably well defined. One is a low potential cytochrome *b* [65, 66], a two-subunit protein which is absent in X-linked chronic granulomatous disease [67]. It is located in both the plasma membrane and specific granule membrane of resting cells, and concentrates in the phagosomal membrane on stimulation. There are good reasons for considering it to be the terminal component of the chain that transfers electrons to O_2 [66], although there is still a body of opinion favouring the involvement of the cytochrome in the oxidase system but not in this capacity [61]. There is also strong evidence for participation of a flavoprotein [12, 61, 68]. Ubiquinone has

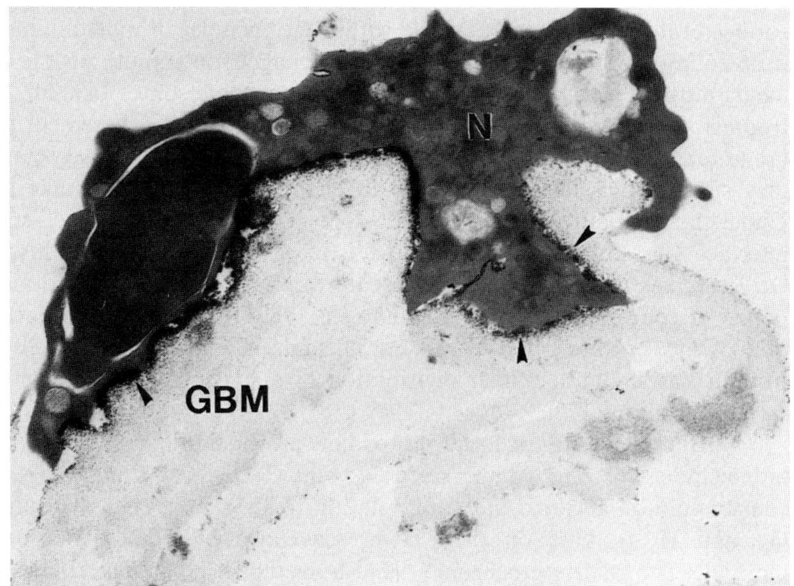

been considered as another oxidase component [69], but this now seems unlikely [62].

In summary, therefore, the pathway of electron transfer in the neutrophil oxidase system is probably from glucose to NADPH via the hexose monophosphate shunt, and then from a flavoprotein to the cytochrome b and finally to O_2.

Quantitation and Sites of Oxidant Production

As discussed above, the NADPH oxidase is located on the neutrophil surface membrane, and can generate oxidants either on the outside of the cell or within cytoplasmic vacuoles or phagosomes. Both O_2^- and H_2O_2 have been measured as products of the reaction, and a number of studies have focused on both the site of production [63, 64, 70–72], and on whether the O_2 consumed by the cells is all converted to O_2^- or whether it can be reduced directly by the enzyme to H_2O_2 [73–78].

Cytochemical techniques have been used to study the distribution of oxidase activity. Most studies have used the method of Briggs et al. [63], in which cerium chloride added to the cells forms an insoluble cerium hydroperoxide precipitate at the site of H_2O_2 production. With soluble stimuli (e.g. PMA, fig. 2a) deposits of the hydroperoxide are seen over the surface of the cells, but particularly within the vacuoles or indentations that form following stimulation [63, 71]. With phagocytic stimuli, on the other hand, most of the deposit is seen within the phagosomes [63, 70]. Neutrophils stimulated by immune complexes embedded in a basement membrane matrix underdo what is commonly termed frustrated phagocytosis – they spread on a surface too large to engulf, and attempt to take it up. Under these conditions [64], peroxide deposit is seen only at the cell-surface interface (fig. 2b), implying that the oxidase is only active in this area. Other studies also suggest that a greater amount of deposit is formed at points of contact between stimulated cells [70, 71]. Cytochemical localization of O_2^- production by PMA- or opsonized zymosan-stimulated neutrophils has shown a similar distribution of activity to that seen with the cerium technique [72].

Further evidence for localized oxidant production comes from quantitative studies relating O_2 uptake to O_2^- and H_2O_2 formation. Whereas O_2 consumption is a global measure, relating to the whole cells, detection of O_2^- and H_2O_2 relies on their being scavenged by other, often larger, molecules (normally cytochrome c and horseradish peroxidase). These will penetrate to an indeterminate extent into vacuoles or phagosomes. Therefore, low recoveries of O_2^- or H_2O_2 could indicate a site of generation not readily accessible to the outside of the cells. A complication is that the net O_2 uptake measured depends on the fate of the O_2^- or H_2O_2 that is generated [74, 76]. As shown by the following equations, different amounts of O_2 are regenerated depending on how they react:

$$
\begin{array}{lll}
O_2^-: & \text{dismutation} & O_2^- + H^+ \rightarrow \tfrac{1}{2}O_2 + \tfrac{1}{2}H_2O_2 \\
& \text{oxidation (e.g. by cyt. } c) & O_2^- + R \rightarrow O_2 + R^{\cdot -} \\
& \text{reduction} & O_2^- + RH_2 + H^+ \rightarrow RH^{\cdot} + H_2O_2 \\
H_2O_2: & \text{catalatic breakdown} & H_2O_2 \rightarrow \tfrac{1}{2}O_2 + \tfrac{1}{2}H_2O \\
& \text{peroxidatic (e.g. with } Cl^-) & H_2O_2 + R^- \rightarrow ROH + OH^-
\end{array}
$$

Taking this into account, some quantitative assessment of the oxidative burst has been made. With soluble stimuli such as PMA, virtually all the O_2 consumed is converted to O_2^- [75], which under appropriate conditions can be recovered as H_2O_2 [74]. This requires blocking peroxidase or catalase activity, otherwise no H_2O_2 accumulates, and even in continuous assays of H_2O_2, intracellular breakdown appears to be a significant factor [73]. The near quantitative recoveries with PMA suggest that the vacuoles seen by electron microscopy as the major sites of oxidant production (fig. 2a) [63, 71] may be mainly channels or indentations. With particulate stimuli, a lower proportion of the O_2 consumed is recovered as extracellular O_2^- or H_2O_2 [73, 74]. However, with the use of cytochalasin B and appropriate inhibitors, yields can be increased, and are also compatible with quantative conversion of O_2 to O_2^- which dismutates to H_2O_2, provided the majority of the O_2^- is formed at an intracellular or intravacuolar location. With neutrophils stimulated by surface-bound immune complexes, none of the O_2 consumed can be recovered as O_2^- and very little as H_2O_2 in the surrounding medium [64]. This strongly supports the cytochemical evidence (fig. 2a) that the oxidants are formed at the area of contact which is inaccessible to the detecting agents.

Neutrophils taking up unopsonized latex, although undergoing a normal burst of O_2 uptake and producing H_2O_2 as expected, yield no detectable O_2^- [61]. While this may mean that all O_2^- production occurs at inaccessible sites [77], it raises the possibility that direct reduction of O_2 to H_2O_2 may occur under some conditions. There is some support for this interpretation in the findings of Green and Wu [78] that the partially purified NADPH-oxidase, depending on substrate concentration, can give varying amounts of one and two electron O_2 reduction.

It makes good sense for the phagocytosing neutrophil to release most of its oxidants, as well as granule enzymes, at its target within the phagocytic vacuole. Oxidant release at a surface is really only an extension of this, but it has wide implications when considering neutrophil-mediated tissue damaging mechanisms. First, with tissue-bound stimuli such as immune complexes, and possibly also with soluble stimuli that cause adherence of the cells, neutrophils are likely to direct their oxidants (and granule enzymes) directly at the tissue [41, 64, 79, 80]. Secondly, inhibitors or scavengers may not gain access to the site of oxidant release, and may therefore be unable to prevent their subsequent reactions.

Quantitative studies have also given insight into the fate of O_2^- and H_2O_2 generated by stimulated neutrophils. Most of the O_2^- appears to dismutate (either spontaneously or catalytically) rather than causing net oxidation or reduction [73]. Catalase activity can account for more than half the H_2O_2 removal, with the remainder undergoing peroxidative reactions [76, 81]. GSH peroxidase plays only a minor role [76], although oxidation of intracellular GSH [82], and of ascorbic acid [83], does occur. Catalatic breakdown could either be by intracellular catalase or (as discussed below) by myeloperoxidase [84].

The respiratory burst, in addition to generating reactive oxygen species, has another important facet. As a result of NADPH oxidation producing H^+ inside the cell, and extracellular dismutation of O_2^- consuming H^+, there is a net translocation of H^+ from the outside to the inside of the cell. Further H^+ would be consumed extracellularly by subsequent HOCl and chloramine formation from the H_2O_2. This almost certainly is the reason why the phagocytic vacuoles of normal neutrophils, but not chronic granulomatous disease cells, undergo an initial alkalinization [85, 86].

It is generally agreed that the production of O_2^- and H_2O_2 alone is insufficient to explain the bactericidal and cytotoxic activity of the neutrophil respiratory burst [60, 61, 87, 88]. O_2^- undergoes very few biologically damaging reactions and, as a radical, it is relatively benign [89]. H_2O_2 is more reactive, although high concentrations are required to be bactericidal [90]. It is considered, therefore, that O_2^- and H_2O_2 must be precursors of more toxic species. The main contenders for this role in the neutrophil are hydroxyl radicals (OH·), hypochlorous acid (HOCl) and chloramines. However, there is still considerable debate about if and how these species are formed, and their importance in the neutrophil armory. These issues are discussed below.

Hydroxyl Radical

There are three intermediates in the reduction of O_2 to H_2O. O_2^-, H_2O_2 and the hydroxyl radical (OH·) are formed by successive one electron additions. OH· is by far the most reactive [89]. It reacts with little discrimination with virtually all biological molecules, and these reactions occur within a narrow radius of its site of generation. Furthermore, many reactions of OH· involve hydrogen abstraction from an organic molecule, thereby generating another radical. This can lead to a chain reaction, amplifying the effect of the initial OH·. Thus OH· certainly has the potential to be a biologically damaging species.

Two approaches have been taken to determine whether or not neutrophils produce OH·. One has been to investigate whether the cells contain the necessary components of an OH· producing reaction, and the other has been to add OH· detecting reagents to stimulated cells. However, as discussed below, neither approach is straightforward, and on critical examination, studies frequently quoted in support of OH· as an important neutrophil oxidant are equivocal. In fact, the bulk of current evidence on neutrophil OH· production is negative.

Are Conditions for OH· Production Met? OH· can be formed from O_2^- and H_2O_2. However, quite restrictive conditions apply. The reaction does not occur spontaneously, but must be catalysed by a transition metal ion [89, 91]. Iron salts are the preferred catalyst, but copper could also participate. The reaction, commonly referred to as either the superoxide-driven Fenton reaction, or the iron catalysed Haber-Weiss reaction, comprises two parts:

$$H_2O_2 + Fe^{2+}(\text{complex}) \rightarrow Fe^{3+}(\text{complex}) + OH^- + OH· \tag{1}$$

$$O_2^- + Fe^{3+}(\text{complex}) \rightarrow Fe^{2+}(\text{complex}) + O_2 \tag{2}$$

For iron to act catalytically, the reduced complex formed in the Fenton reaction 1 must be recycled. O_2^- is not specifically required for this. Reductants such as ascorbic acid, thiols or organic reducing radicals, e.g. semiquinones, are alternatives. In the stimulated neutrophil, however, O_2^- is best placed to serve this function.

For the Haber-Weiss reaction to proceed, not only must iron (or its equivalent) be present, but its complexed form is critical. Although Fe(EDTA) is an excellent catalyst of the reaction, other iron complexes are not [92, 93]. Fe(citrate), Fe(ADP), Fe(ATP) and a number of other potential physiological chelates all give scarcely detectable OH· production. Hence the existence of physiological low molecular weight iron catalysts of the Haber-Weiss reaction remains to be established. It is possible, however, that iron in the absence of a chelating agent might participate in a mechanism similar to reactions 1 and 2, but producing an oxidant which does not behave as free OH·. This would be analogous to what is seen with xanthine oxidase-derived O_2^-, although not with O_2^- generated radiolytically [94], when the oxidant appears to be an iron-peroxide complex or ferryl species [95]. However, its relevance to the neutrophil O_2^--generating enzyme is unknown.

To produce OH·, therefore, the neutrophil requires iron in a catalytically active form. Stimulated cells release two iron proteins, myeloperoxidase and lactoferrin. However, myeloperoxidase, discussed below, is not considered to produce OH·, and although it has been reported that lactoferrin is a good Haber-Weiss catalyst [96], subsequent studies have not confirmed these findings [93, 97, 98]. They have found it catalytically inactive, either when relatively unsaturated or with both iron-binding sites occupied. Contrary results may be explained by the presence of EDTA, which confers on lactoferrin some catalytic activity [97]. Further, measurements of neutrophil iron binding capacity [99], and the iron content of lactoferrin extracted from neutrophil granules [Vissers and Winterbourn, unpublished observation] suggest it is <25% saturated, and current evidence favours it functioning more as an inhibitor of iron-dependent reactions.

An alternative to an endogenous Haber-Weiss catalyst would be exogenous iron or iron taken up by the neutrophil during phagocytosis. Iron in plasma or extracellular fluids exists almost exclusively tightly bound to protein. Although it has been reported that rheumatoid synovial fluid contains micromolar concentrations of readily chelated iron [100], we have not been able to measure such iron in fluids examined in our laboratory. Potential host-derived iron sources are, therefore, transferrin, haemoglobin and other haem proteins, and ferritin. Transferrin, like lactoferrin, has been shown to be catalytically inactive [93, 98]. Haem proteins, including haemoglobin, do react with H_2O_2, but in this respect they function as peroxidases, capable of oxidizing or hydroxylating many biological compounds, but not generating OH˙ [101]. Haem compounds, therefore, although not Haber-Weiss catalysts, in combination with neutrophil H_2O_2 could enhance bactericidal or tissue-damaging activity. Ferritin, however, may be able to release catalytic iron [91, 102, 103]. Recent studies have shown that either an O_2^--generating system [91], or stimulated neutrophils [103], can mobilize ferritin iron, but whether this leads to OH˙ production by the neutrophils remains to be seen. The fate or availability of phagocytosed bacterial iron is not known. Repine et al. [104] have shown that iron loading increases bacterial susceptibility to H_2O_2, and implicated an OH˙-dependent mechanism. However, they found that the loaded bacteria were killed normally by neutrophils [105].

Taken together, therefore, these studies lead to the conclusion that conditions in the neutrophil are not generally favourable for OH˙ formation.

Do Stimulated Neutrophils Produce OH˙? A number of studies have shown that stimulated neutrophils can oxidize methional [106], α-keto-γ-methiolbutyric acid (KMB) [107] or dimethylsulphoxide (DMSO) [108]. While these are all reactions undergone by OH˙, there are several reasons why they are not conclusive evidence for OH˙ production by the cells. Firstly, the detection reactions are not necessarily specific. They are a better indicator of OH˙ involvement if appropriate scavengers inhibit in accordance with their OH˙ reaction rate constants, but in no case was this observed with all the scavengers studied. Benzoate, ethanol and mannitol tended to be weak inhibitors, and the strongest inhibitor was thiourea, which is notably poor at discriminating OH˙ [109]. Furthermore, the observed inhibition by cyanide [107, 108] suggests involvement of myeloperoxidase. There is little methional oxidation by myeloperoxidase-deficient cells [110]. Secondly, although the reactions were inhibited by superoxide dismutase, in most cases catalase was reported as having very little effect. This is not compatible with a Haber-Weiss mechanism, or any other mechanism of OH˙ production based on the Fenton reaction.

Thirdly, all the studies were carried out without rigorously removing adventitious iron that would have been present in buffers or media. The possibility that this iron was involved as a catalyst cannot be excluded. Addressing this question, Thomas et al. [111] have recently shown that only with exogenous iron present do neutrophils peroxidize polyunsaturated lipid attached to phagocytosed beads, and have concluded that without this iron the cells do not produce strongly oxidizing radicals such as OH˙.

Others have used the spin trap, 5, 5-dimethyl-1-pyrroline-N-oxide (DMPO), and taken formation of the hydroxyl adduct (DMPO-OH) as evidence of OH· formation [112, 113]. The above criticisms also apply to these studies, with further complications inherent to the spin-trapping technique. As discussed by Rosen and Finkelstein [114], DMPO-OH can be formed either by direct trapping of OH· or via breakdown of the O_2^- adduct. With this in mind, Britigan et al. [115] have reinvestigated the spin-trapped radicals of stimulated neutrophils, and have concluded that trapped O_2^- is the source of DMPO-OH, and that, at least without a substantial exogenous iron source, OH· is not a significant product of the cells.

Hence, although the results of earlier studies are equivocal regarding OH· being a significant oxidant product of stimulated neutrophils, two recent investigations [111, 115] provide convincing evidence that this is not the case. The possibility remains that OH· could be formed if the cells are presented with an appropriate catalyst [116], but such a catalyst is yet to be identified.

Even if neutrophils did have access to Haber-Weiss catalyst, another mechanism operates against their producing OH·. This is the consumption of H_2O_2 by myeloperoxidase [117]. As discussed below, myeloperoxidase can either react with H_2O_2 and Cl^- to form hypochlorous acid (HOCl) or break down H_2O_2 through its catalase activity. Either way it diverts H_2O_2 from other reactions such as OH· production. Myeloperoxidase at 10 nM is able to inhibit fully OH· production from O_2^- and H_2O_2 catalysed by 2 μM Fe(EDTA), implying that OH· production would only occur in the phagosome or neutrophil surroundings if a catalyst were present in considerable excess over myeloperoxidase [117]. Since myeloperoxidase is by far the most abundant protein released on degranulation, this would not readily be achieved.

In spite of the foregoing doubts that neutrophils can convert their O_2^- and H_2O_2 to OH·, several in vivo or whole tissue studies have implicated OH· in neutrophil-mediated cell damage. The best documented example is the lung damage seen in association with complement activation [118–120], where protection by iron chelators such as desferrioxamine or lactoferrin and by some OH· scavengers was indicative of OH· involvement. However, this interpretation leaves a number of unanswered questions. For example, it is unclear why some but not other scavengers are protective; how addition of another OH· scavenger to a biological medium such as plasma, which is already a 'scavenger soup', improves its protective ability; and why exogenous iron chelators are effective in the presence of endogenous lactoferrin or transferrin. Clearly a greater understanding of the chemistry of these complex physiological processes is required, and at this stage it seems premature to conclude that oxidant damage is due to OH·.

Myeloperoxidase and Hypochlorous Acid

Myeloperoxidase is released from the azurophil granules of stimulated neutrophils. As first described by Klebanoff and co-workers, the isolated enzyme, in combination with H_2O_2 and a halide ion, readily kills bacteria and fungi [88, 90, 121]. The original studies were with I^-, and showed iodination of microbial constituents, but even though the order of reactivity is $I^- > Br^- > Cl^-$, under physiological conditions Cl^- is the preferred substrate. The mechanism of conversion of Cl^- to chlorinating species has been studied by a number of authors since the original work of Agner [122], and the product of the reaction has been identified as HOCl [123]. The purified enzyme exhibits complex kinetics. Its pH optimum is approximately 5, where HOCl is thought to be formed in a 2-electron oxidation via a classical peroxidase mechanism:

$$\text{(ferric myeloperoxidase)} \quad MP^{3+} + H_2O_2 \rightarrow MP^{3+}H_2O_2 \text{ (compound I)}$$

$$MP^{3+}H_2O_2 + Cl^- \rightarrow MP^{3+} + OH^- + HOCl$$

Both H_2O_2 and Cl^- inhibit HOCl formation at high concentrations. Inhibition by Cl^- requires binding to a protonated site and, therefore, is important at lower pH [124, 125]. This binding also increases the K_m for H_2O_2. Thus at pH 5 the enzyme functions best with low Cl^- and high H_2O_2. At pH 7, where the rate of HOCl formation is much less, the reaction is favoured by high Cl^- and low H_2O_2 concentrations.

Myeloperoxidase can also act as a catalase. The proportion of H_2O_2 broken down catalatically compared with that converted to HOCl increases with increasing H_2O_2 concentration [84]. Thus at pH 7.6, HOCl is formed stoichiometrically with up to approximately 100 μM H_2O_2 but millimolar H_2O_2 is almost all broken down catalatically to O_2 and H_2O. Therefore, depending on the pH, Cl^- and H_2O_2 concentrations in the environment where it is released, the neutrophil enzyme could function either as a peroxidase or a catalase.

Although the majority of mechanistic studies with myeloperoxidase have been carried out near its pH optimum, it now appears that in the neutrophil, most of its action is likely to occur at neutral pH. The pH of the phagosome (and presumably the frustrated phagocytosis equivalent) initially rises slightly due to NADPH oxidase activity and remains near neutral for several minutes before becoming acidic [85, 86]. During this time most of the O_2^- reduction occurs, and killing takes place. In addition, neutrophils produce much less O_2^- or H_2O_2 at pH 6 than at pH 7.4 [126]. Hence, myeloperoxidase released either extracellularly or into phagosomes should have maximum substrate availability at neutral pH. This appears paradoxical, considering the pH optimum for chlorination. Furthermore, during phagocytosis, most of the O_2^- and H_2O_2 are released into the phagosome in a short burst. This is likely to generate a high concentration of H_2O_2 in the narrow space between the phagosome membrane and engulfed object. This concentration is not known, but by our estimate it could be greater than millimolar [84]. If so, myeloperoxidase should break down most of the H_2O_2 as a catalase, and only when the H_2O_2 concentration (and possibly pH) dropped, would it convert the residual H_2O_2 to HOCl.

Another conundrum regarding the function of myeloperoxidase is that on neutrophil stimulation it is converted by co-released O_2^- to the superoxide adduct (compound III). Characteristic peaks can be seen in the difference spectrum of stimulated versus resting cells which shifts to the spectrum of ferrous myeloperoxidase when the cells become anaerobic [84]. Compound III, like oxyhaemoglobin, releases O_2 reversibly on deoxygenation [127]. Compound III is well characterized as the product of the reaction between O_2^- and myeloperoxidase [128], but in terms of the classical peroxidase mechanism, it is considered to be an inactive side product and its formation should prevent HOCl production by the cells [129]. Yet findings that myeloperoxidase is converted to compound III by either a xanthine oxidase O_2^-/H_2O_2 generating system, or by PMA-stimulated neutrophils, but still chlorinates monochlorodimedon, imply it is catalytically active [84]. In these studies, superoxide dismutase, which prevented this conversion, did not affect chlorination. We have recently confirmed that conversion to compound III has little effect on chlorination at low rates of H_2O_2/O_2^- generation. However, at high rates, superoxide dismutase inhibits the reaction [130]. This inhibition parallels an increase in compound II formation, discernible in the myeloperoxidase spectrum. Hence O_2^- accelerates the reaction in an analogous way to low concentrations of ascorbate [130], by preventing accumulation of compound II. As a consequence, O_2^- reverses inhibition by H_2O_2, and gives greater chlorination at high H_2O_2 concentrations [130]. This effect of O_2^- may be an important factor in enabling myeloperoxidase to chlorinate more efficiently within the phagosome.

H_2O_2 is required for chlorination involving compound III [84], but otherwise, the mechanism is not established. The situation at pH 7.8 appears different from that at low pH where superoxide dismutase gives a three- to fourfold increase in chlorination [130]. Myeloperoxidase also appears able to break down O_2^- catalytically. Although superoxide dismutase activity has been suggested [132], we find no catalytic O_2^- breakdown in the presence of catalase, and propose that myeloperoxidase can act as a combined catalase/superoxide dismutase [130].

Evidence that neutrophils produce HOCl has been obtained by several groups [84, 133–135]. It is too reactive to be a stable product, however, and must be trapped with excess of a reagent such as monochlordimedon, taurine or methionine. Conversion of up to approximately 50% of the theoretical O_2 consumed has been achieved, using soluble stimuli such as PMA [134]. Under such conditions, most of the H_2O_2 is released to the exterior of the cell so high concentrations are unlikely to build up [136]. The extent to which HOCl is formed during phagocytosis is less well established. There is evidence that some chlorination of ingested particles occurs [133,135] but there is very little HOCl-dependent bleaching of phagocytosed fluorescein-coated beads [85, 86]. Nauseef et al. [76], in accounting for the fate of O_2 consumed by neutrophils ingesting opsonized *S. aureus*, concluded that up to 70% of H_2O_2 formed was decomposed to O_2 and H_2O, with a lesser fraction undergoing peroxidative reactions. Although they did not identify the source of catalatic activity, it seems reasonable to conclude that at least some was due to myeloperoxidase rather than intracellular catalase and, therefore, that myeloperoxidase functions both as a catalase and peroxidase during phagocytosis.

Myeloperoxidase can undergo peroxidative reactions independently of Cl^-. Physiological Cl^- concentrations favour HOCl formation, but whether other peroxidative reactions could be important in neutrophil function has not been explored. Another area of possible significance is intragranular reactions, as it has been observed that H_2O_2 added to resting neutrophils can to some extent escape intracellular decomposition and react with myeloperoxidase located in the azurophil granules [137]. The significance of such intragranular peroxidative reactions is not known.

In summary, therefore, current evidence suggests that myeloperoxidase is converted to the superoxide adduct (compound III) when released from stimulated neutrophils. It acts mainly at neutral pH, where compound III can participate both in the formation of HOCl and catalatic breakdown of H_2O_2. Within the neutrophil phagosome, high H_2O_2 concentrations during the early stages of stimulation would favour myeloperoxidase functioning mainly as a catalase, although O_2^- may enhance its chlorinating ability. As the H_2O_2 concentration and pH decline, chlorination should become more effective. Extracellularly, where the H_2O_2 concentration remains low, HOCl formation should predominate.

Hypochlorous Acid and Chloramines

HOCl is extremely reactive. Either added as a reagent or generated from myeloperoxidase, it is readily microbicidal or cytotoxic [9, 88, 138–

140]. It reacts with practically all biological compounds, including haem proteins and porphyrins, thiols, thioethers, ascorbic acid, reduced pyridine nucleotides, amines and amino acids [139–143]. It is unlikely, therefore, to survive beyond the immediate vicinity of the neutrophil, and its reactions will depend on relative concentrations and relative reactivities of compounds in this region. Thiol groups and methionine are about 100 times more reactive than other amino acids or amines, e.g. taurine, and appear most vulnerable [141]. However, the variety of reactive groups in any biological preparation means that HOCl is unlikely to modify one component specifically. Likewise, without a large excess of HOCl, inactivation of any one component is unlikely to be complete.

The reaction of HOCl with thiols forms a mixture of the disulphide and higher oxidation products [141], methionine forms the sulphoxide [142], and amines and amino acids form N-chloramines [139, 143–146]. Those formed from α-amino acids are unstable and rapidly decompose to give aldehydes, which may help perpetuate the original oxidative damage. Amines such as taurine and ammonia form more stable chloramines, while retaining the oxidizing capacity of the HOCl.

Current interest in chloramines stems from the proposition that these longer-lived oxidants may increase the selectivity and extend the range of the neutrophil myeloperoxidase system. They are not as reactive as HOCl but, nevertheless, can oxidize thiols, thioethers and haem proteins [147]. Lipophilic chloramines such as ammonia-derived NH_2Cl readily kill bacteria, but hydrophilic compounds such as taurine chloramine are only weakly bactericidal [145]. Taurine protects against killing by HOCl or the myeloperoxidase-H_2O_2-Cl^- system [148, 149]. Whereas HOCl reacts exclusively with the cell surface, different processes appear to be involved in chloramine toxicity [149].

Stimulated neutrophils have been shown to produce stable chloramines in vitro. After PMA stimulation, 10% of the myeloperoxidase-derived oxidants can be recovered in the medium at the end of the experiment as endogenous chloramines [135, 150]. Neutrophils contain appreciable concentrations of taurine, some of which appears to be secreted during stimulation [151], and taurine chloramine is the major endogenous chloramine formed. Significant amounts of NH_2Cl are also detectable [152, 153]. Whether chloramines are formed in appreciable quantities in vivo will depend on environmental factors. Amine concentrations must be sufficient to compete favourably with other functional groups on ingested targets or in the extracellular fluid that react with HOCl, and further study is required to determine the significance of chloramines as physiological neutrophil oxidants.

Singlet Oxygen

Singlet oxygen ($^1O_2$), the electronically excited state of molecular oxygen, still receives attention in some current literature as a likely product of the neutrophil oxidative burst. It was originally proposed as the source of chemiluminescence generated by stimulated cells [154]. However, subsequent studies have shown that traps commonly used to scavenge $^1O_2$ also react with HOCl, and chemiluminescence is due to less specific oxidations [153, 156, and reviewed in ref. 61]. Foote et al. [157] found virtually none of the characteristic $^1O_2$ oxidation product following neutrophil ingestion of cholesterol attached to phagocytosable beads. Furthermore, although the isolated myeloperoxidase system can generate $^1O_2$, optimal conditions are way outside the physiological range, and the reaction is unlikely to be relevant to neutrophil function [158]. With this body of negative evidence, it now seems appropriate to exclude $^1O_2$, from lists of potential neutrophil oxidants.

Chemiluminescence, nevertheless, is a regular accompaniment of the respiratory burst, and is frequently used to monitor neutrophil stimulation. The chemiluminescent products have not been identified, but are presumed to be transient oxidation products of neutrophil components and the stimulus. Both myeloperoxidase-dependent and independent processes contribute to the natural chemiluminescence of stimulated cells [159]. When used as an analytical technique, chemiluminescence is usually measured in the presence of luminol to amplify the response. In this case, it is almost entirely due to myeloperoxidase-dependent reactions [160].

Lipid Peroxidation

Lipid peroxidation is a well-documented free radical chain reaction, which can he initiated by OH· or transition metal complexes [89]. Yet few studies have focused on its involvement in neutrophil oxidative reactions. Peroxidation of phagocytosed lipid was first observed by Stossel et al. [161]. Carlin [162] found a dependence on iron but, surprisingly, no inhibition by superoxide dismutase or catalase. The most recent studies of Thomas et al. [111] showed that there is no lipid peroxidation without exogenously added iron. This reinforces the point made for OH· production, that positive results may be obtained only because of adventitiously present transition metal ions. Moreover, if lipid peroxidation is physiologically significant, it becomes important to identify sources of catalyst available to the neutrophil.

Lipid peroxidation products have been detected in phagocytosed bacteria [161], and in red cells surrounding stimulated neutrophils [163]. Iron involvement in these reactions has not been characterized, although with the red cells, haemoglobin did not appear to participate. Recent studies by Sepe and Clark [148, 164] have shown that the myeloperoxidase system, and stimulated neutrophils, can lyse phospholipid liposomes. Although a peroxidation mechanism is suggested by the enhancement by polysaturated lipid, and protection by α-tocopherol or β-carotene, this was not tested directly, and the authors suggested a chlorination mechanism. A more extensive investigation of neutrophil-mediated peroxidation, and the interactions between lipid and the myeloperoxidase system, is warranted.

Lipid peroxidation occurring as an 'uncontrolled' process, and generally viewed in terms of cell damage, should be distinguished from 'controlled' peroxidation, as in lipoxygenase-catalysed conversion of arachidonic acid to leukotrienes. Stimulation of neutrophils causes leukotriene production, so must be associated with at least some lipid peroxidation.

Microbial Killing

Neutrophils rapidly kill phagocytosed micro-organisms, as well as muticellular parasites too large to ingest. Intraphagosomal killing is accomplished within a few minutes. The respiratory burst is required for killing most, but not all micro-organisms, and defective killing is seen under anaerobic conditions or in chronic granulomatous disease [6–9, 87, 149, 165]. Furthermore, many of the bacteria killed by chronic granulomatous disease cells produce H_2O_2 [166], suggesting that oxidative processes utilizing this H_2O_2 could still account for much of the microbicidal activity of these cells. Neutrophils also have granule constituents which are capable of killing without a requirement for O_2 [149, 165]. They include cationic proteins related to chymotrypsin, a bacterial permeability-increasing protein [167] and a group of 3,000 to 5,000-kD peptides termed defensins [168]. All these are most effective at neutral pH and act by binding to the surface of the target. Another bactericidal protein, associated with azurophil granule membranes and with an acid pH optimum, has recently been described [169]. The molecular basis for any of these non-oxidative bacterial killing mechanisms is not established.

From studies with isolated systems, it would appear that myeloperoxidase-derived HOCl, and possibly lipophilic chloramines, are the neutrophil oxidants best suited for microbial killing. O_2^- alone is not considered toxic, and most micro-organisms are sensitive only to high concentrations of H_2O_2 [90, 121]. $OH^\cdot$ kills readily [104, 170] but, as discussed above, it is not a proven component of the neutrophil armoury. It is not at all clear, however, what role any one microbicidal system, either oxidative or non-oxidative, plays in the neutrophil itself. This partly stems from the likelihood that the cells do not rely on a single mechanism for carrying out their major function. Different mechanisms may also operate for different classes of organisms. Killing can also occur both intraphagosomally and extracellularly. Scavenger studies have revealed a role for oxidants, and for myeloperoxidase, in extracellular killing [168, 171, 172]. Similar studies of intracellular killing are hampered by the logistical problem of introducing sufficient scavenger into the phagosome. Johnston et al. [6] showed that superoxide dismutase or catalase attached to latex beads inhibited killing, but other studies have been less conclusive. Bacterial superoxide dismutase or catalase content does not correlate with resistance to neutrophil killing [173], but this could be explained by killing being a rapid, extracellular event [153, 174] and not influenced by intracellular constituents. The best evidence for myeloperoxidase involvement in intraphagosomal killing is inhibition by azide or cyanide [175, 176]. However, there is the conundrum that severe myeloperoxidase deficiency is common (one in several thousand [177]) and very seldom associated clinically with recurrent infections [177, 178]. In vitro assays have shown that *Candida* species are killed poorly by myeloperoxidase-deficient cells [179], killing of *S. aureus* is slightly defective [175, 176, 179], but *E. coli* appear to be killed normally [174]. With *E. coli*, subsequent inactivation of bacterial enzymes was defective for the deficient neutrophils, suggesting that myeloperoxidase may be involved not in the initial killing, but in the inactivation and breakdown of microbial components [174]. It appears, therefore, that the myeloperoxidase system may be more significant in killing some organisms than others. If it is generally important in normal cells, other backup mechanisms must take over in its absence. Certainly H_2O_2 can build up in myeloperoxidase-deficient cells, and it may act directly, or perhaps by a mechanism involving O_2^-.

Segal [87] has suggested a different reason why the respiratory burst might be required for microbicidal activity. He has postulated that it is a source of alkalinity rather than oxidants, and maintains an initial phagosomal environment in which neutral-acting enzymes and microbicidal proteins can act. In this scheme, myeloperoxidase would act, at least initially, predominantly to remove O_2^- and H_2O_2 formed as a byproduct of the alkalinization reaction. This could be compatible with its kinetic properties [84], and might also explain why myeloperoxidase is not necessary for microbicidal activity and may not be involved in the initial killing step [174]. This mechanism could be combined with HOCl being produced when the H_2O_2 concentration and pH of the environment became more favourable, to assist in degradation and possibly (as discussed below) to down-regulate the whole neutrophil response.

Cytotoxicity

Neutrophils are capable of damaging and killing host cells. Such action could be directed against specific cells, as in antibody-dependent cytotoxicity, or host cells could be innocent bystanders in an inflammatory response. A considerable body of recent literature documents neutrophil-mediated injury to isolated or cultured cells [reviewed in ref. 180], and in vivo studies imply that stimulation of neutrophils by inflammatory mediators results in tissue injury where the cells accumulate [118–120, 181–183].

Similar mechanisms are likely to operate in host cell killing as in microbicidal activity, although there is an inherent difference when target cells are not the prime stimulus, and not phagocytosed. However, stimulated neutrophils become adhesive, particularly to vascular endothelium, and are then likely to act while spread over the target cell surface [10, 41, 79, 182, 184]. Both oxidative and non-oxidative injurious mechanisms have been identified. The operative mechanism may depend on the nature of the target cells and whether they are phagocytosed, and particularly on what parameter is used to monitor cell injury.

Lysis of red blood cells by stimulated neutrophils has been shown to be oxidative and myeloperoxidase-dependent [185–187]. H_2O_2 and chloramines can enter the cells and cause haemoglobin oxidation [185, 186], but it appears to be HOCl, acting at the cell surface, which is cytolytic [187]. Neutrophils stimulated independently with opsonized zymosan or PMA [188–191] can also lyse various tumour cell lines. Again, in most cases, myeloperoxidase-derived oxidants were responsible for cytolysis. The mechanism of antibody-dependent lysis, when the cells are phagocytosed, is less definitive [192–194]. Although most studies have shown oxidant involvement, there is no clear requirement for myeloperoxidase. H_2O_2 alone, however, can also be cytotoxic [195–196], and it has recently been reported that anti-microbial defensins, isolated from neutrophil granules, can kill tumour cells [197]. Endothelial cells are also injured by stimulated neutrophils. Again, lysis appears to be oxidant-mediated [198–201], although in the study by Weiss et al. [199], myeloperoxidase involvement could not be demonstrated.

Whereas lysis is the ultimate, gross measure of cytotoxicity, tissue injury can be manifested by other less extreme cellular changes. Cell adhesion and the ability to maintain a permeability barrier have been studied with both endothelial and epithelial cells, and impaired function observed under conditions where lysis was not apparent [202–205]. Neutral proteinase activity has been implicated in endothelial cell detachment [205], but permeability changes have been attributed to both oxidative [202, 204, 206] and non-oxidative [203, 204] mechanisms.

Even more subtle changes affecting cell metabolism can arise from exposure to stimulated neutrophils. Cultured endothelial cells or pneumocytes have decreased ATP levels, decreased protein and DNA synthesis, and release prostacyclin [207, 208]. Platelet aggregation can be inhibited [209]. H_2O_2 produced by the neutrophils has been implicated and, in fact, most of these changes can be mimicked by exposure of the cells to H_2O_2 or an enzymatic H_2O_2-generating system [207–210]. Sublethal H_2O_2 concentrations can also stimulate K^+ efflux and purine release by endothelial cells [211] and inhibit proteoglycan and hyaluronic acid synthesis by chondrocytes [212, 213]. This suggests that neutrophil H_2O_2 production may be important in modulating host cell metabolism at inflammatory sites. This may lead to impaired cell function or to the release of other mediators, for example, prostaglandins, which extend the inflammatory response. The latter supports the concept proposed by Lands [214] that peroxide levels could provide a control mechanism over inflammatory reactions.

In a number of systems, susceptibility to lysis by neutrophils has been shown to depend on the anti-oxidant defences of the target cells [191, 215, 216]. Removal of H_2O_2 appears to be the major factor. The susceptibility of some tumour cell lines and endothelial cells to H_2O_2 is enhanced by blocking GSH metabolism [215–217], but with other tumour cells [218], and red blood cells [219, 220], catalase offers the greatest protection. Relative susceptibilities of different tumour cell lines, at least to H_2O_2, cannot be fully explained in terms of their levels of anti-oxidant enzymes, however [218].

Red cells, as a consequence of their ability to dispose of H_2O_2, can also be considered as a buffer against oxidative reactions occurring in the blood stream [136, 186, 219, 220]. They can inhibit extracellular reactions of H_2O_2, and the formation of HOCl from stimulated neutrophils [220], and they also appear to be protective in models of lung injury [221, 222].

Connective Tissue Degradation

There is good evidence that neutrophils collecting at inflammatory sites contribute to the breakdown of surrounding connective tissue. This could be an important part of the remodelling process, but out of control it could lead to excessive destruction, as observed in chronic inflammatory conditions such as rheumatoid arthritis. Stimulated neutrophils release a number of proteinases capable of degrading different types of collagen, and structural proteins such as fibronectin, laminin and cartilage proteoglycan [79, 181, 223, 224]. Neutrophil oxidants may also attack matrix macromolecules.

O_2^- and H_2O_2 generating systems, or radiolytically generated $OH^{\cdot}$ have been shown to depolymerize isolated or synovial fluid hyaluronic acid [225–228], to inhibit collagen gelation [229], and to degrade soluble collagen or cartilage-proteoglycan aggregates [230–232]. Hyaluronic acid degradation has also been observed with stimulated neutrophils. However, these reactions were observed when iron (and usually EDTA) was present, and can all be attributed to a strong oxidant of such as $OH^{\cdot}$. Thus the physiological relevance of degradation of the type seen in these systems hinges on the ability of neutrophils to generate $OH^{\cdot}$. The ability of myeloperoxidase-derived oxidants to degrade structural macromolecules directly has not been tested.

As distinct from direct degradation, neutrophil oxidants can modify proteins so as to increase their susceptibility to proteolytic enzymes. Increased proteolysis of glomerular basement membrane proteins or albumin pretreated with myeloperoxidase-H_2O_2-Cl^- [224], or albumin exposed to $OH^{\cdot}$ radicals [233] has been observed. This mechanism has recently been considered for neutrophils degrading either extracellular matrix laid down by endothelial cells [234], or isolated (insoluble) glomerular basement membrane [224]. In both cases, all the protein degradation measured (hydroxyproline solubilization from type IV collagen) was proteolytic. Although an effect of oxidants was not seen when expressed on a per cell basis, one became apparent when basement membrane degradation was related to the levels of active proteinase released by the neutrophils [224]. Cells deficient in oxidant production (chronic granulomatous disease or myeloperoxidase deficient) or treated with oxidant scavengers released considerably more active proteinase, but caused no more degradation than untreated cells. The implication from these findings is that myeloperoxidase-derived oxidants can not only inactivate neutrophil proteinases (see below), but also enhance the digestibility of their substrate. Overall, however, the two effects compensate for one another. Neutrophil oxidants influence structural protein degradation, therefore, but their effect would appear to be secondary to that of co-released proteinases. However, more subtle oxidative changes may occur, altering functional properties such as cell adhesion or subunit assembly of matrix constituents.

Regulatory and Suicidal Activity of Neutrophil Oxidants

Neutrophils are not immune to their own toxicity mechanisms. Stimulated cells autoinactivate [235], and intracellular anti-oxidant defenses, particularly the glutathione system, can protect against some of this inactivation [236]. On first impression, this suicidal action may seem wasteful. However, considering it is necessary to terminate as well as initiate host defence responses, autoinactivation may be an important regulatory mechanism. Moreover, there are now numerous examples of oxidative inactivation and, in some cases, activation of mediators of neutrophil function and other host-defense reactions, suggesting a general modulatory role for oxidants in the inflammatory response. As detailed below, myeloperoxidase or HOCl has almost invariably been implicated in these effects.

Compared with normal cells myeloperoxidase-deficient neutrophils exhibit enhanced phagocytosis, their respiratory burst although initially normal extends for longer periods, and they release more active granule enzymes [77, 224, 237–241]. Myeloperoxidase-dependent oxidative inactivation of various components of the stimulus-response mechanism appears to be largely responsible. The myeloperoxidase system has been shown to inactivate the NADPH oxidase [239, 240]. It also inactivates or down-regulates many phagocytic or soluble stimuli or their receptors [241], including complement factors C_{5a} and C_3, the Fc region of immune complexes, leukotriene C_4 and the chemotactic peptide fMet-Leu-Phe [237, 242, 243]. Both these mechanisms could act in concert to terminate the response. Voetman et al. [244] first described the oxidative inactivation of lysozyme and β-glucuronidase during release from stimulated neutrophils, and subsequent studies have shown that vitamin B_{12} binding protein [245] and the neutral proteinases elastase, cathepsin G, gelatinase and collagenase [224, 246] also undergo myeloperoxidase-dependent inactivation. As shown in figure 3, some granule enzymes are more readily inactivated than others.

Enzyme activation by neutrophil oxidants has also been reported. Gelatinase and collagenase are metalloproteinases stored in neutrophil granules in a latent form, and both can be partially activated by HOCl or the myeloperoxidase system [247–249]. However, these observations must be viewed in the context that neutral proteinases released from neutrophils readily activate gelatinase [250], and that collagenase and gelatinase are both susceptible to oxidative inactivation [246].

Other inflammatory mediators inactivated by the neutrophil myeloperoxidase system include a number of microbial toxins. Agner [251] first described the inactivation of diphtheria and tetanus toxins, and similar findings have been reported for pneumolysin, *Actinobacillus* leukotoxin and *Clostridium* cytotoxin [252, 253]. Myeloperoxidase can also inhibit lymphocyte proliferation and natural killer cell activity [254]. However, although stimulated neutrophils have been shown to down-regulate natural killer cells, in one study proteinases [255], and in the other, oxidants but not myeloperoxidase [256], were implicated.

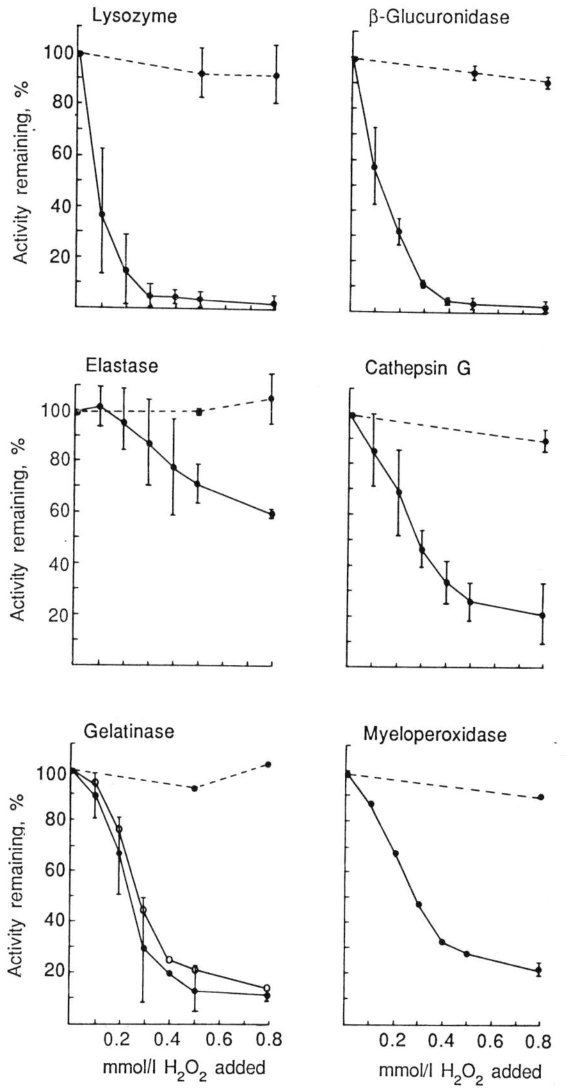

Fig. 3. Inactivation of exocytosed neutrophil granule enzymes. Serial additions of 100 nmol H_2O_2 were added to the supernatant of stimulated neutrophils. Dotted lines represent similar experiments carried out in the presence of azide (or in the case of myeloperoxidase, catalase). The two solid lines for gelatinase represent two assay methods. Results taken from ref. [246] (with permission from Biochem J).

A reaction possibly involved in controlling neutrophil-mediated degradation of host tissue in inflammation is the myeloperoxidase-dependent inactivation of α_1-proteinase inhibitor [147, 237, 257, 258]. α_1-Proteinase inhibitor is the major inhibitor of elastase in plasma or extracellular fluids, and it is proposed that its inactivation would enable released elastase to remain active in the vicinity of stimulated neutrophils [259]. Beyond the range of myeloperoxidase action, however, unoxidized inhibitor would still complex elastase and restrict its area of influence.

Taken together, these findings do suggest that neutrophil oxidants (in particular myeloperoxidase-derived HOCl) could be involved in regulating inflammatory responses. Nevertheless, such observations must be interpreted cautiously. Firstly, because of its extreme reactivity, HOCl will modify the function of many biological compounds. Hence oxidative modification will not be specific for regulatory molecules. Secondly, although inactivation may occur readily with purified systems, it will be much less facile in physiological situations where each compound will have to compete with numerous others for reaction with the available HOCl. Thus, only with a very large excess of HOCl would substantial inactivation of any one component occur.

Summary and Conclusions

Neutrophils produce reduced oxygen species in response to a wide range of particulate and soluble inflammatory stimuli. Other mediators activate the cells to enhance their stimulus response. The O_2-reducing enzyme is an electron-transport chain within the cell membrane. Depending on the stimulus, O_2 reduction can occur on the exterior of the cell, within phagosomes, or at the area of contact between the cell and a stimulatory surface. The respiratory burst produces mainly (if not exclusively) O_2^-, which can dismutase to give H_2O_2. Evidence for OH$^\cdot$ production is at best equivocal, and conditions provided by the cells are not well suited for OH$^\cdot$ production by conventional mechanisms. Myeloperoxidase released from neutrophil granules can combine with H_2O_2 and Cl$^-$ to form HOCl, which is potently microbicidal and cytotoxic. However, myeloperoxidase can also break down O_2^- and act as catalase, and when released by neutrophils it is converted to its superoxide adduct (compound III). Within phagosomes or extracellularly, myeloperoxidase should act mainly at neutral pH, in spite of its pH 5–6 optimum. Under these conditions, compound III, like the native enzyme, catalyses HOCl formation, provided the H_2O_2 concentration remains low, but both convert to catalase activity as the H_2O_2 increases towards millimolar. The significance of all these reactions in the neutrophil is not yet clear.

Oxidant production is a key requirement for efficient microbial killing. Yet direct evidence for the involvement of a particular oxidant in the process remains elusive. The myeloperoxidase system appears particularly suited to this role, but myeloperoxidase deficiency is common and largely asymptomatic, and is associated with impaired in vitro killing of only some micro-organisms. The cells also contain granule constituents with non-oxidative microbicidal activity, but their participation in neutrophil killing is also poorly understood. Neutrophils are capable of injuring host cells as well as invading pathogens. Cytolysis of a variety of cell types has been observed, and is mainly myeloperoxidase-dependent, whereas both oxidative and non-oxidative processes can modify key properties, such as adhesion and maintenance of a permeability barrier. More subtle neutrophil-mediated changes, due to direct effects of H_2O_2 and affecting the metabolism of target or bystander cells, have also been observed. These cellular effects, plus the ability of myeloperoxidase-derived oxidants to enhance the proteolytic susceptibility of connective tissue proteins are all ways in which neutrophil oxidants could contribute to inflammatory damage to host tissue. Myeloperoxidase-derived oxidants, through inactivation of neutrophil enzymes and inflammatory mediators, may also play a role in regulating neutrophil function and terminating their inflammatory action.

Acknowledgements

I am grateful to John French and Tony Kettle for helpful comments on the manuscript, and to Sophie Harrison Sands for expert secretarial assistance.

References

1   Baldridge CW, Gerard RW: The extra respiration of phagocytes. Am J Physiol 1933;103:235–236.
2   Sbarra AJ, Karnovsky ML: The biochemical basis of phagocytosis. 1. Metabolic changes during the ingestion of particles by polymorphonuclear leukocytes. J Biol Chem 1959;234:1355–1362.
3   Iyer GYN, Islam MF, Quastel JH: Biochemical aspects of phagocytosis. Nature 1961;192:535–541.
4   Babior BM, Kipnes RS, Curnutte JT: Biological defense mechanisms: the production by leukocytes of superoxide, a potential bactericidal agent. J Clin Invest 1973;52:741–744.
5   Drath DB, Karnovsky ML: Superoxide production by phagocytic leukocytes. J Exp Med 1975;141:257–262.
6   Johnston RB, Keele BB Jr, Misra HP, et al: The role of superoxide anion generation in phagocytic bacteridical activity. Studies with normal and chronic granulomatous disease leukocytes. J Clin Invest 1975; 55:1357–1372.

7 Mandell GL: Bactericidal activity of aerobic and anaerobic polymorphonuclear neutrophils. J Clin Invest 1972;51:2566–2572.

8 Holmes B, Page AR, Good RA: Studies of the metabolic activity of leukocytes from patients with a genetic abnormality of phagocytic function. J Clin Invest 1967;46:1422–1432.

9 Babior BM: Oxidant-dependent microbial killing. N Engl J Med 1978;298:659–668,721–725.

10 Weissman G, Smolen JE, Korchak HM: Release of inflammatory mediators from stimulated neutrophils. N Engl J Med 1980;303:27–34.

11 Baggiolini M, Dewald B: Exocytosis by neutrophils; in Snyderman R (ed): Regulation of Leukocyte Function. New York, Plenum Press, 1984, pp 221–246.

12 Rossi F: The O_2^--forming NADPH oxidase of the phagocytes: nature, mechanisms of activation and function. Biochim Biophys Acta 1986;853:65–89.

13 Repine JE, White JG, Clawson CC, et al: The influence of phorbol myristate acetate on oxygen consumption by polymorphonuclear leukocytes. J Lab Clin Med 1974;83:911–920.

14 Becker EL, Showell HJ, Henson PM, et al: The ability of chemotactic factors to induce lysosomal enzyme release. J Immunol 1974;112:2047–2054.

15 Goldstein IM, Roos D, Kaplan HB, et al: Complement and immunoglobulins stimulate superoxide production by human leukocytes independently of phagocytosis. J Clin Invest 1975;56:1155–1163.

16 Cohen HJ, Chovaniec ME, Wilson MK, et al: Con-A-stimulated superoxide production by granulocytes: Reversible activation of NADPH oxidase. Blood 1982;60:1188–1194.

17 Cohen HJ, Chovaniec ME, Takahashi K, et al: Activation of human granulocytes by arachidonic acid: Its use and limitations for investigating granulocyte functions. Blood 1986;67:1103–1109.

18 Serhan CN, Radin A, Smolen JE, et al: Leukotriene B_4 is a complete secretagogue in human neutrophils: A kinetic analysis. Biochem Biophys Res Commun 1982;107:1006–1012.

19 Tzeng DY, Deuel TF, Huang JS, et al: Platelet-derived growth factor promotes polymorphonuclear leukocyte activation. Blood 1984;64:1123–1128.

20 Tsujimoto M, Yokota S, Vilcek J, et al: Tumor necrosis factor provokes superoxide anion generation from neutrophils. Biochem Biophys Res Commun 1986;137:1094–1100.

21 Dewald B, Baggiolini M: Platelet-activating factor as a stimulus of exocytosis in human neutrophils. Biochim Biophys Acta 1986;888:42–48.

22 Nishizuka Y: The role of protein kinase C in cell surface signal transduction and tumor promotion. Nature 1984;308:693–698.

23 Berridge MJ: Calcium. A universal second messenger. Triangle 1985; 24:79–90.

24 Snyderman R (ed): Contemporary Topics of Immunobiology: Regulation of Leukocyte Function. New York, Plenum Press, 1984, vol 14.

25 Tauber AI: Protein kinase C and activation of the human neutrophil NADPH-oxidase. Blood 1987;69:711–720.

26 Papini E, Grzeskowiak M, Bellavite P, et al: Protein kinase C phosphorylates a component of NADPH oxidase of neutrophils. FEBS Lett 1985;190:204–208.

27 Heyworth PG, Segal AW: Further evidence for the involvement of a phosphoprotein in the respiratory burst oxidase of human neutrophils. Biochem J 1986;239:723–731.

28 McPhail LC, Snyderman R: Activation of the respiratory burst enzyme in human polymorphonuclear leukocytes by chemoattractants and other soluble stimuli: evidence that the same oxidase is activated by different transductional enzymes. J Clin Invest 1983;72:192–200.

29 Cooke E, Hallett MB: The role of C-kinase in the physiological activation of the neutrophil oxidase. Evidence from using pharmacological manipulation of C-kinase activity in intact cells. Biochem J 1985;232:323–327.

30 Gerard C, McPhail LC, Marfat A, et al: Role of protein kinases in stimulation of human polymorphonuclear leukocyte oxidative metabolism by various agonists. Differential effects of a novel protein kinase inhibitor. J Clin Invest 1986;77:61–65.

31 O'Flaherty JT, Schmitt JD, McCall CE, et al: Diacylglycerols enhance human neutrophil degranulation responses: relevancy to a multiple mediator hypothesis of cell function. Biochem Biophys Res Commun 1984;123:64–70.

32 Johnston RB Jr, Lehmeyer JE: Elaboration of toxic oxygen by-products by neutrophils in a model of immune complex disease. J Clin Invest 1976;57:836–841.

33 Henson PM, Hollister JR, Musson RA, et al: Inflammation as a surface phenomenon: initiation of inflammatory processes by surface-bound immunologic components; in Weissman G, et al (eds): Advances in Inflammation Research. New York, Raven Press, 1979, vol 1, pp 341–352.

34 Takemura R, Werb Z: Secretory products of macrophages and their physiological functions. Am J Physiol 1984;246(Cell Physiol 15):C1–C9.

35 Lopez AF, Williamson DJ, Gamble JR, et al: Recombinant human granulocyte-macrophage colony-stimulating factor stimulates in vitro mature human neutrophil and eosinophil function, surface receptor expression, and survival. J Clin Invest 1986;78:1220–1228.

36 Weisbart RH, Golde DW, Clarke SC, et al: Human granulocyte-macrophage colony-stimulating factor is a neutrophil activator. Nature 1985;314:361–363,

37 Shalaby MR, Aggarwal BB, Rinderknect E, et al: Activation of human polymorphonuclear neutrophil functions by interferon-γ and tumor necrosis factors. J Immunol 1985;135:2069–2073.

38 Klebanoff SJ, Vadas MA, Harlan JM, et al: Stimulation of neutrophils by tumor necrosis factor. J Immunol 1986;136:4220–4225.

39 Dewald B, Baggiolini M: Activation of NADPH oxidase in human neutrophils. Synergism between fMLP and the neutrophil products PAF and LTB_4. Biochem Biophys Res Commun 1985;128:297–304.

40 Guthrie LA, McPhail LC, Henson PM, et al: Priming of neutrophils for enhanced release of oxygen metabolites by bacterial lipopolysaccharide. Evidence for increased activity of the superoxide-producing enzyme. J Exp Med 1984;160:1656–1671.

41 Worthen GS, Haslett C, Smedly LA, et al: Lung vascular injury induced by chemotactic factors: enhancement by bacterial endotoxins. Fed Proc 1986;45:7–12.

42 Smedly LA, Tonnesen MG, Sandhaus RA, et al: Neutrophil-mediated injury to endothelial cells. Enhancement by endotoxin and essential role of neutrophil elastase. J Clin Invest 1986;77:1233–1243.

43 Van Epps DE, Garcia ML: Enhancement of neutrophil function as a result of prior exposure to chemotactic factor. J Clin Invest 1980;66:167–175.

44 English D, Roloff JS, Lukens JN: Chemotactic factor enhancement of superoxide release from fluoride and phorbol myristate acetate stimulated neutrophils. Blood 1981;58:129–134.

45 French JK, Hurst NP, Zalewski PD, et al: Calcium ionophore A23187 enhances human neutrophil superoxide release, stimulated by phorbol dibutyrate, by converting phorbol ester receptors from a low- to high-affinity state. FEBS Lett 1987;212:242–246.

46 Bender JG, McPhail LC, Van Epps DE: Exposure of human neutrophils to chemotactic factors potentiates activation of the respirtory burst enzyme. J Immunol 1983;130:2316–2323.

47 McCall CE, Bass DA, DeChatelet LR, et al: In vitro responses of human neutrophils to N-formyl-methionyl-leucyl-phenylalanine: Correlation with effects of acute bacterial infection. J Infect Dis 1979;140:277–286.

48 Bass DA, Olbrantz P, Szejda P, et al: Subpopulations of neutrophils with increased oxidative product formation in blood of patients with infection. J Immunol 1986;136:860–866.

49 Smith RJ, Iden SS: Pharmacological modulation of chemotactic factor-elicited release of granule-associated enzymes from human neutrophils. Effects of prostaglandins, non-steroid anti-inflammatory agents and corticosteroids. Biochem Pharmacol 1980;29:2389–2393.

50 Kaplan HB, Edelson HS, Korchak HM, et al: Effects of non-steroidal anti-inflammatory agents on human neutrophil functions in vitro and in vivo. Biochem Pharmacol 1984;33:371–378.

51 Gay JC, Lukens JN, English DK: Differential inhibition of neutrophil superoxide generation by nonsteroidal anti-inflammatory drugs. Inflammation 1984;8:209–222.

52 Neal TM, Vissers MCM, Winterbourn CC: Inhibition by nonsteroidal anti-inflammatory drugs of superoxide production and granule enzyme release by polymorphonuclear leukocytes stimulated with immune complexes and formyl-methionyl-leucyl-phenylalanine. Biochem Pharmacol 1987;36:2511–2517.

53 Neal TM, Vissers MCM, Winterbourn CC: Inhibition of neutrophil-mediated degradation of isolated basement membrane collagen by nonsteroidal anti-inflammatory drugs that inhibit degranulation. Arthritis Rheum 1987;30:908–913.

54 Minta JO, Williams MD: Some nonsteroidal anti-inflammatory drugs inhibit the generation of superoxide anions by activated polymorphs by blocking ligand-receptor interactions. J Rheumatol 1985;12:751–757.

55 Venezio FR, DiVincenzo C, Pearlman F, et al: Effects of the newer nonsteroidal anti-inflammatory agents, ibuprofen, fenoprofen, and sulindac, on neutrophil adherence. J Infect Dis 1985;152:690–694.

56 Perianin A, Torres M, Labro M-T, et al: The different inhibitory effects of phenylbutazone on soluble and particle stimulation of human neutrophil oxidative burst. Biochem Pharmacol 1983;32:2819–2822.

57 Hammershmidt DE, White JG, Craddock PR, et al: Corticosteroids inhibit complement-induced granulocyte aggregation: A possible mechanism for their efficacy in shock states. J Clin Invest 1979;63:798–803.

58 Jacob HS: Complement-mediated leucoembolization: A mechanism of tissue damage during extracorporeal perfusions, myocardial infarction and in shock–A review. Quart J Med 1983;LII:289–296.

59 Rossi F, Zatti M: Biochemical aspects of phagocytosis in polymorphonuclear leukocytes. NADH and NADPH oxidation by the granules of resting and phagocytozing cells. Experientia 1964;20:21–23.

60 Badwey JA, Karnovsky ML: Active oxygen species and the functions of phagocytotic leukocytes. Annu Rev Biochem 1980;48:695–726.

61 Tauber AI, Babior BM: Neutrophil oxygen reduction: the enzymes and the products. Adv Free Radicals Biol Med 1985;1:265–308.

62 Cross AR, Jones OTG, Garcia R, et al: The subcellular localization of ubiquinone in human neutrophils. Biochem J 1983;216:765–768.

63 Briggs, RT, Drath DB, Karnovsky ML, et al: Localization of NADH oxidase on the surface of human polymorphonuclear leukocytes by a new cytochemical method. J Cell Biol 1975;67:566–586.

64 Vissers MCM, Day AW, Winterbourn CC: Neutrophils adherent to a non-phagocy-

tosable surface (glomerular basement membrane) produce oxidants only at the site of attachment. Blood 1985;66:161–166.

65 Segal AW, Jones OTG: Novel cytochrome b systems in phagocytic vacuoles of human granulocytes. Nature 1978;276:515–517.

66 Cross, AR, Parkinson JF, Jones OTG: Mechanism of the superoxide-producing oxidase of neutrophils. O_2 is necessary for the fast reduction of cytochrome b-245 by NADPH. Biochem J 1985;226:881–884.

67 Segal AW: Absence of both cytochrome b-245 subunits from neutrophils in X-linked chronic granulomatous disease. Nature 1987;326:88–91.

68 Babior BM, Kipnes RS: Superoxide-forming enzyme from human neutrophils: Evidence for a flavin requirement. Blood 1977;50:517–524.

69 Crawford DR, Schneider DL: Identification of ubiquinone-50 in human neutrophils and its role in microbicidal events. J Biol Chem 1982;257:6662–6668.

70 Ohno Y-I, Hirai K-I, Kanoh T, et al: Subcellular localization of H_2O_2 production in human neutrophils stimulated with particles and an effect of cytochalasin-B on the cells. Blood 1982;60:253–260.

71 Ohno Y-I, Hirai K-I, Kanoh T, et al: Subcellular localization of hydrogen peroxide production in human polymorphonuclear leukocytes stimulated with lectins, phorbol myristate acetate, and digitonin: An electron microscopic study using $CeCl_3$. Blood 1982;60:1195–1202.

72 Briggs RT, Robinson JM, Karnovsky ML, et al: Superoxide production by polymorphonuclear leukocytes. A cytochemical approach. Histochemistry 1986;84:371–378.

73 Root RK, Metcalf JA: H_2O_2 release from human granulocytes during phagocytosis. Relationship to superoxide anion formation and cellular catabolism of H_2O_2: Studies with normal and cytochalasin B-treated cells. J Clin Invest 1977;60:1266–1279.

74 Dri P, Bellavite P, Berton G, et al: Interrelationship between oxygen consumption, superoxide anion and hydrogen peroxide formation in phagocytosing guinea pig polymorphonuclear leucocytes. Mol Cell Biochem 1979;23:109–122.

75 Makino R, Tanaka T, Iizuka T, et al: Stoichiometric conversion of oxygen to superoxide anion during the respiratory burst in neutrophils. J Biol Chem 1986;261:11444–11447.

76 Nauseef WM, Metcalf JA, Root RK: Role of myeloperoxidase in the respiratory burst of human neutrophils. Blood 1983;61:483–492.

77 Hallett MB, Campbell AK: Two distinct mechanisms for stimulation of oxygen-radical production by polymorphonuclear leucocytes. Biochem J 216:459–465.

78 Green, TR, Wu DE: The NADPH:O_2 oxidoreductase of human neutrophils. Stoichiometry of univalent and divalent reduction of O_2. J Biol Chem 1986;261:6010–6015.

79 Henson PM, Hollister JR, Musson RA, et al: Inflammation as a surface phenomenon: initiation of inflammatory process by surface bound immunologic components. Adv Inflamm Res 1979;1:341–352.

80 Vissers MCM, Winterbourn CC, Hunt JS: Degradation of glomerular basement membrane by human neutrophils in vitro. Biochim Biophys Acta 1984;804:154–160.

81 Borregaard N, Schwartz JH, Tauber AI: Proton secretion by stimulated neutrophils. Significance of hexose monophosphate shunt activity as source of electrons and protons for the respiratory burst. J Clin Invest 1984;74:455–459.

82 Voetman AA, Loos JA, Roos D: Changes in the levels of glutathione in phagocytosing human neutrophils. Blood 1980;55:741–747.

83 Winterbourn CC, Vissers MCM: Changes in ascorbate levels on stimulation of human neutrophils. Biochim Biophys Acta 1983;763:175–179.

84 Winterbourn CC, Garcia R, Segal AW: Production of the superoxide adduct of

myeloperoxidase (compound III) by stimulated neutrophils, and its reactivity with H_2O_2 and chloride. Biochem J 1985;228:583–592.

85 Segal AW, Geisow M, Garcia R, et al: The respiratory burst of phagocytic cells is associated with a rise in vacuolar pH. Nature 1981;290:406–409.

86 Cech P, Lehrer RI: Phagolysosomal pH of human neutrophils. Blood 1984;36:88–95.

87 Segal AW: Variations on the theme of chronic granulomatous disease. Lancet 1985;i:1378–1382.

88 Klebanoff SJ: Oxygen metabolism and the toxic properties of phagocytes. An Intern Med 1980;93:480–489.

89 Halliwell B, Gutteridge JMC: Oxygen toxicity, oxygen radicals, transition metals and disease. Biochem J 1984;219:1–14.

90 McRipley RJ, Sbarra AJ: Role of the phagocyte in host-parasite interactions. XII. Hydrogen peroxide myeloperoxidase bacterial system in the phagocyte. J Bacteriol 1967;94:1425–1430.

91 Aust SD, Morehouse LA, Thomas CE: Role of metals in oxygen radical reactions. J Free Radicals Biol Med 1985;1:3–26.

92 Sutton HC: Efficiency of chelated iron compounds as catalysts for the Haber-Weiss reaction. J Free Radicals Biol Med 1985;1:195–202.

93 Baker MS, Gebicki JM: The effect of pH on hydroxyl radicals produced from superoxide by potential biological iron chelators. Arch Biochem Biophys 1986;246:581–588.

94 Winterbourn CC, Sutton HC: Iron and xanthine oxidase catalyse formation of an oxidant species distinguishable from OH·: comparison with the Haber-Weiss reaction. Arch Biochem Biophys 1986;244:27–36.

95 Sutton HC, Vile GF, Winterbourn CC: Radical driven Fenton reactions–Evidence from paraquat radical studies for production of tetravalent iron in the presence and absence of EDTA. Arch Biochem Biophys 1987;256:462–471.

96 Ambruso DR, Johnston RB: Lactoferrin enhances hydroxyl radical production by human neutrophils, neutrophil particulate fractions, and an enzymatic generating system. J Clin Invest 1981;67:352–360.

97 Winterbourn CC: Lactoferrin-catalysed hydroxyl radical production: additional requirement for a chelating agent. Biochem J 1983;210:15–19.

98 Baldwin DA, Jenny ER, Aisen P: The effect of human serum transferrin and milk lactoferrin on hydroxyl radical formation from superoxide and hydrogen peroxide. J Biol Chem 1984;259:13391–13394.

99 Van Snick JL, Masson PL, Heremans JF: The involvement of lactoferrin in the hyposideremia of acute inflammation. J Exp Med 1974;140:1068–1084.

100 Gutteridge JMC, Rowley DA, Halliwell B: Superoxide-dependent formation of hydroxyl radicals and lipid peroxidation in the presence of iron salts. Biochem J 1982;206:605–609.

101 Winterbourn CC: Free radical production and oxidative reactions of hemoglobin. Environ Health Perspect 1985;64:321–330.

102 O'Connell M, Halliwell B, Moorhouse CP, et al: Formation of hydroxyl radicals in the presence of ferritin and haemisiderin. Biochem J 1986;234:727–731.

103 Biemond P, van Eijk HG, Swaak AJG, et al: Iron mobilization from ferritin by superoxide derived from stimulated polymorphonuclear leukocytes. Possible mechanism in inflammation disease. J Clin Invest 1984;73:1576–1579.

104 Repine JE, Fox RB, Berger EM: Hydrogen peroxide kills *Staphylococcus aureus* by reacting with staphylococcal iron to form hydroxyl radical. J Biol Chem 1981;256:7094–7096.

105 Repine JE, Fox RB, Berger EM, et al: Effect of staphylococcal iron content on the killing of *Staphylococcus aureus* by polymorphonuclear leukocytes. Infect Immun 1981;32:407–410.

106 Tauber AI, Babior BM: Evidence for hydroxyl radical production by human neutrophils. J Clin Invest 1977;60:374–379.

107 Weiss SJ, Rustagi PK, LoBuglio AF: Human granulocyte generation of hydroxyl radical. J Exp Med 1978;147:316–323.

108 Repine JE, Eaton JW, Anders MW, et al: Generation of hydroxyl radical by enzymes, chemicals, and human phagocytes in vitro. J Clin Invest 1979;64:1642–1651.

109 Winterbourn CC: The ability of scavengers to distinguish OH· production in the iron-catalysed Haber-Weiss reaction: Comparison of four assays for OH·. J Free Radicals Biol Med 1987;3:33–39.

110 Klebanoff SJ, Rosen H: Ethylene formation by polymorphonuclear leukocytes: Role of myeloperoxidase. J Exp Med 1978;148:490–506.

111 Thomas MJ, Shirley PS, Hedrick CC, et al: Role of free radical processes in stimulated human polymorphonuclear leukocytes. Biochemistry 1986;25:8042–8048.

112 Green MR, Hill HAO, Okolow-Zubkowska MJ, et al: The production of hydroxyl and superoxide radicals by stimulated human neutrophils–Measurements by EPR spectroscopy. FEBS Lett 1979;100:23–26.

113 Rosen H, Klebanoff SJ: Hydroxyl radical generation by polymorphonuclear leukocytes measured by electron spin resonance spectroscopy. J Clin Invest 1979;64:1725–1729.

114 Rosen GM, Finkelstein E: Use of spin traps in biological systems. Adv Free Radical Biol Med 1985;1:345–375.

115 Britigan BE, Rosen GM, Chai Y, et al: Do human neutrophils make hydroxyl radical? Determination of free radicals generated by human neutrophils activated with a soluble or particulate stimulus using electron paramagnetic resonance spectrometry. J Biol Chem 1986;261:4426–4431.

116 Halliwell B, Gutteridge JMC, Blake D: Metal ions and oxygen radical reactions in human inflammatory joint disease. Philos Trans R Soc Lond 1985;B311:659–671.

117 Winterbourn CC: Myeloperoxidase as an effective inhibitor of hydroxyl radical production: Implications of the oxidative reactions of neutrophils. J Clin Invest 1986;78:545–550.

118 Ward PA, Till GO, Kunkel R, et al: Evidence for role of hydroxyl radical in complement and neutrophil-dependent tissue injury. J Clin Invest 1983;72:789–801.

119 Till GO, Hatherill JR, Tourtellotte WW, et al: Lipid peroxidation and acute lung injury after thermal trauma to skin. Evidence of a role for hydroxyl radical. Am J Pathol 1985;119:376–384.

120 Ward PA, Till GO, Hatherill JR, et al: Systemic complement activation, lung injury, and products of lipid peroxidation. J Clin Invest 1985;76:517–527.

121 Klebanoff SJ: Iodination of bacteria: A bactericidal mechanism. J Exp Med 1967;126:1063–1078.

122 Agner K: Verdoperoxidase. A ferment isolated from leukocytes. Acta Physiol Scand 1941;2:(suppl 8):1–62.

123 Harrison JE, Schultz J: Studies on the chlorinating activity of myeloperoxidase. J Biol Chem 1976;251:1371–1374.

124 Andrews PC, Krinsky NI: A kinetic analysis of the interaction of human myeloperoxidase with hydrogen peroxide, chloride ions, and protons. J Biol Chem 1982;257:13240–13245.

125 Bakkenist ARJ, De Boer JEG, Plat H, et al: The halide complexes of myeloperoxidase and the mechanism of the halogenation reactions. Biochim Biophys Acta 1980;613:337–348.

126 Gabig TG, Bearman SI, Babior BM: Effects of oxygen tension and pH on the respiratory burst of human neutrophils. Blood 1979;53:1133–1139.

127 Odajima T, Yamazaki I: Myeloperoxidase of the leukocyte of normal blood. I. Reaction of myeloperoxidase with hydrogen peroxide. Biochim Biophys Acta 1970;206:71–77.

128 Odajima T, Yamazaki I: Myeloperoxidase of the leukocyte of normal blood. III. The reaction of ferric myeloperoxidase with superoxide anion. Biochim Biophys Acta 1972;284:355–359.

129 Yamazaki I, Yokota K: Oxidation states of peroxidase. Mol Cell Biochem 1973;2:39–52.

130 Kettle AJ, Winterbourn CC: Superoxide modulates the activity of myeloperoxidase and optimizes the production of bypochlorous acid. Biochem J 1988; 252:529–536.

131 Bolscher BGJM, Zoutberg GR, Cuperus RA, et al: Vitamin C stimulates the chlorinating activity of human myeloperoxidase. Biochim Biophys Acta 1984;784:189–191.

132 Cuperus RA, Muijsers AO, Wever R: The superoxide dismutase activity of myeloperoxidase; formation of compound III. Biochim Biophys Acta 1986;871:78–84.

133 Foote CS, Goyne TE, Lehrer RE: Assessment of chlorination by human neutrophils. Nature 1983;301:715–716.

134 Weiss SJ, Klein R, Slivka A, et al: Chlorination of taurine by human neutrophils. J Clin Invest 1982;70:598–607.

135 Adeniyi-Jones SK, Karnovsky ML: Oxidative decarboxylation of free and peptide-linked amino acids in phagocytozing guinea pig granulocytes. J Clin Invest 1981;68:365–373.

136 Test ST, Weiss SJ: Quantitative and temporal characterization of the extracellular H_2O_2 pool generated by human neutrophils. J Biol Chem 1984;259:399–405.

137 Ohno Y, Gallin JI: Diffusion of extracellular hydrogen peroxide into intracellular compartments of human neutrophils. Studies utilizing the inactivation of myeloperoxidase by hydrogen peroxide and azide. J Biol Chem 1985;260:8438–8446.

138 Albrich JM, Hurst JK: Oxidative inactivation of *Escherichia coli* by hypochlorous acid. Rates and differentiation of respiratory from other reaction sites. FEBS Lett 1982;144:157–161.

139 Thomas EL: Myeloperoxidase, hydrogen peroxide, chloride antimicrobial system: Nitrogen-chlorine derivatives of bacterial components in bactericidal action against *Escherichia coli*. Infect Immun 1979;23:522–531.

140 Albrich JM, McCarthy CA, Hurst JK: Biological reactivity of hypochlorous acid: Implications for microbicidal mechanisms of leukocyte myeloperoxidase. Proc Natl Acad Sci USA 1981;78:210–214.

141 Winterbourn CC: Comparative reactivities of various biological compounds with myeloperoxidase-hydrogen peroxide-chloride, and similarity of the oxidant to hypochlorite. Biochim Biophys Acta 1985;840:204–210.

142 Tsan M-F: Myeloperoxidase-mediated oxidation of methionine and amino acid decarboxylation. Infect Immun 1982;36:136–141.

143 Zgliczynski JM, Stelamaszynska T, Domanski J, et al: Chloramines as intermediates of oxidation reaction of amino acids by myeloperoxidase. Biochim Biophys Acta 1971;235:419–424.

144 Stelamaszynska T, Zgliczynski JM: N-(2-oxoacyl) amino acids and nitriles as final products of dipeptide chlorination mediated by the myeloperoxidase/H_2O_2/Cl^- system. Eur J Biochem 1978;92:301–308.

145 Thomas EL: Myeloperoxidase-hydrogen peroxide-chloride antimicrobial system: effect of exogenous amines on antibacterial action against *Escherichia coli*. Infect Immun 1979;25:110–116.

146 Weiss SJ, Lampert MB, Test ST: Long-lived oxidants generated by human neutrophils: Characterization and bioactivity. Science 1983;222:625–628.

147 Test ST, Weiss, SJ: The generation and utilization of chlorinated oxidants by human neutrophils. Adv Free Radical Biol Med 1986;2:91–116.

148 Sepe SM, Clark RA: Oxidant membrane injury by the neutrophil myeloperoxidase system. II. Injury by stimulated neutrophils and protection by lipid-soluble antioxidants. J Immunol 1985;134:1896–1901.

149 Root RK, Cohen MS: The microbicidal mechanisms of human neutrophils and eosinophils. Rev Infect Dis 1981;3:565–598.

150 Test ST, Lampert MB, Ossana JG, et al: Generation of nitrogen-chlorine oxidants by human phagocytes. J Clin Invest 1984;74:1341–1349.

151 Thomas EL, Grisham MB, Jefferson MM: Myeloperoxidase-dependent effect of amines on functions of isolated neutrophils. J Clin Invest 1983;72:441–454.

152 Grisham MB, Jefferson MM, Melton DF, et al: Chlorination of endogenous amines by isolated neutrophils. Ammonia-dependent bactericidal, cytotoxic, and cytolytic activities of the chloramines. J Biol Chem 1984;259:10404–10413.

153 Albrich JM, Gilbaugh JH III, Callahan KB, et al: Effects of the putative neutrophil-generated toxin, hypochlorous acid, on membrane permeability and transport systems of *Escherichia coli*. J Clin Invest 1986;78:177–184.

154 Allen RC, Stjernholm RL, Steele RH: Evidence for the generation of an electronic excitation state(s) in human polymorphonuclear leukocytes and its participation in bactericidal activity. Biochem Biophys Res Commun 1972;47:679–684.

155 Harrison JE, Watson BD, Schultz J: Myeloperoxidase and singlet oxygen: A reappraisal. FEBS Lett 1978;92:327–331.

156 Held AM, Hurst JK: Ambiguity associated with use of singlet oxygen trapping agents in myeloperoxidase-catalyzed oxidations. Biochem Biophys Res Commun 1978;81:878–885.

157 Foote CS, Abakerli RB, Clough RL, et al: On the question of singlet oxygen production in leukocytes, macrophages, and the dismutation of superoxide anion; in Bannister WH, Bannister JV (eds): Biological and Clinical Aspects of Superoxide and Superoxide Dismutase. Amsterdam, Elsevier/North Holland, 1980 pp 222–230.

158 Kanofsky JR, Wright J, Miles-Richardson GE, et al: Biochemical requirements for singlet oxygen production by purified human myeloperoxidase. J Clin Invest 1984;74:1489–1495.

159 Cheson BD, Christensen RL, Sperling R, et al: The origin of the chemiluminescence of phagocytosing granulocytes. J Clin Invest 1976;58:789–796.

160 DeChatelet LR, Long GD, Shirley PS, et al: Mechanism of the luminol-dependent chemiluminescence of human neutrophils. J Immunol 1982;129:1589–1593.

161 Stossel TP, Mason RJ, Smith AL: Lipid peroxidation by human blood phagocytes. J Clin Invest 1974;54:638–645.

162 Carlin G: Peroxidaton of linolenic acid promoted by human polymorphonuclear leukocytes. J Free Radicals Biol Med 1985;1:255–262.

163 Claster S, Chiu DT-Y, Quintanilha A, et al: Neutrophils mediate lipid peroxidation in human red cells. Blood 1984;64:1079–1084.

164 Sepe SM, Clark RA: Oxidant membrane injury by the neutrophil myeloperoxidase system. I. Characterization of a liposome model and injury by myeloperoxidase, hydrogen peroxide, and halides. J Immunol 1985;134:1888–1895.

165 Elsbach P, Weiss J: A reevaluation of the roles of the O_2-dependent and O_2-independent microbicidal systems of phagocytes. Rev Infect Dis 1983;5:854–864.

166 Mandell GL, Hook EW: Leukocyte bactericidal activity in chronic granulomatous disease: correlation of bacterial hydrogen peroxide production and susceptibility ot intracellular killing. J Bacteriol 1969;100:531–532.

167 Weiss J, Kao L, Victor M, et al: Oxygen-independent intracellular and oxygen-dependent extracellular killing of *Escherichia coli* S15 by human polymorphonuclear leukocytes. J Clin Invest 1985;76:206–212.

168 Ganz T, Selsted ME, Szklarek D, et al: Defensins. Natural peptide anitbiotics of human neutrophils. J Clin Invest 1985;76:1427–1435.

169 Gabay JE, Heiple JM, Cohn ZA, et al: Subcellular location and properties of bactericidal factors from human neutrophils. J Exp Med 1986;164:1407–1421.

170 Rosen H, Klebanoff SJ: Role of iron and ethylenediaminetetraacetic acid in the bactericial activity of a superoxide anion-generating system. Arch Biochem Biophys 1981;208:512–519.

171 Passo SA, Weiss SJ: Oxidative mechanisms utilized by human neutrophils to destroy *Escherichia coli*. Blood 1984;63:1361–1368.

172 Wagner DK, Collins-Lech C, Sohnle PG: Inhibition of neutrophil killing of *Candida albicans* pseudohyphae by substances which quench hypochlorous acid and chloramines. Infect Immun 1986;51:731–735.

173 Schwartz CE, Krall J, Norton L, et al: Catalase and superoxide dismutase in *Escherichia coli*. Roles in resistance to killing by neutrophils. J Biol Chem 1983;258:6277–6281.

174 Hamers MN, Bot AAM, Weening RS, et al: Kinetics and mechanism of the bactericidal action of human neutrophils against *Escherichia coli*. Blood 1984;64:635–641.

175 Klebanoff SJ: Myeloperoxidase: Contribution to the microbicidal activity of intact leukocytes. Science 1970;169:1095–1097.

176 Miyasaki KT, Wilson ME, Brunetti AJ, et al: Oxidative and nonoxidative killing of *Actinobacillus actinomycetemcomitans* by human neutrophils. Infect Immun 1986;53:154–160.

177 Parry MF, Root RK, Metcalf JA, et al: Myeloperoxidase deficiency. Prevalence and clinical significance. Ann Intern Med 1981;95:293–301.

178 Kitahara M, Eyre HJ, Simonian Y, et al: Hereditary myeloperoxidase deficiency. Blood 1981;57:888–893.

179 Lehrer RI, Cline MJ: Leukocyte myeloperoxidase deficiency and disseminated candidiasis: the role of myeloperoxidase in resistance to *Candida* infection. J Clin Invest 1969;48:1478–1488.

180 Weiss SJ, LoBuglio AF: Biology of disease. Phagocyte-generated oxygen metabolites and cellular injury. Lab Invest 1982;47:5–18.

181 Cochrane CG: Role of granulocytes in immune complex-induced tissue injuries. Inflammation 1977;2:319–333.

182 Craddock PR, Hammerschmidt DE, Moldow CF, et al: Granulocyte aggregation as a manifestation of membrane interactions with complement: Possible role in leukocyte margination, microvascular occlusion, and endothelial damage. Semin Hematol 1979;16:140–147.

183 Shasby DM, Vanbenthuysen KM, Tate RM, et al: Granulocytes mediate acute edematous lung injury in rabbits and isolated rabbit lungs perfused with phorbol myristate acetate: role of oxygen radicals. Am Rev Respir Dis 1982;125:443–447.

184 Harlan JM: Leukocyte-endothelial interactions. Blood 1985;65:513–525.

185 Grisham MB, Jefferson MM, Thomas EL: Role of monochloramine in the oxidation of erythrocyte hemoglobin by stimulated neutrophils. J Biol Chem 1984;259:6757–6765.

186 Thomas EL, Grisham MB, Melton DF, et al: Evidence for a role of taurine in the in vitro oxidative toxicity of neutrophils toward erythrocytes. J Biol Chem 1985;260:3321–3329.

187 Dallegri F, Ballestrero A, Frumento G, et al: Role of hypochlorous acid and chloramines in the extracellular cytolysis by neutrophil polymorphonuclear leukocytes. J Clin Lab Immunol 1986;20:37–41.

188 Clark RA, Klebanoff SJ: Neutrophil-mediated tumor cell cytotoxicity: Role of the peroxidase system. J Exp Med 1975;141:1442–1447.

189 Clark RA, Szot S: The myeloperoxidase-hydrogen peroxide-halide system as effector of neutrophil-mediated tumor cell cytotoxicity. J Immunol 1981;126:1295–1301.

190 Weiss SJ, Slivka A: Monocyte and granulocyte-mediated tumor cell destruction. A role for the hydrogen peroxide-myeloperoxidase-chloride system. J Clin Invest 1982;69:255–262.

191 Dallegri F, Frumento G, Patrone F: Mechanisms of tumor cell destruction by PMA-activated human neutrophils. Immunology 1983;48:273–279.

192 Clark RA, Klebanoff SJ: Studies on the mechanism of anitbody-dependent polymorphonuclear leukocyte-mediated cytotoxicity. J Immunol 1977;119:1413–1418.

193 Katz P, Simone CB, Henkart PA, et al: Mechanisms of anitbody-dependent cellular cytotoxicity. The use of effector cells from chronic granulomatous disease patients as investigative probes. J Clin Invest 1980;65:55–63.

194 Borregaard N, Kragballe K: Role of oxygen in antibody-dependent cytotoxicity mediated by monocytes and neutrophils. J Clin Invest 1980;66:676–683.

195 Hoffmann ME, Meneghini R: Action of hydrogen peroxide on human fibroblasts in culture. Photochem Photobiol 1979;30:151–155.

196 Simon RH, Scoggin CH, Patterson D: Hydrogen peroxide causes the fatal injury to human fibroblasts exposed to oxygen radicals. J Biol Chem 1981;256:7181–7186.

197 Lichtenstein A, Ganz T, Selsted ME, et al: In vitro tumor cell cytolysis mediated by peptide defensins of human and rabbit granulocytes. Blood 1986;68:1407–1410.

198 Sacks T, Moldow CF, Craddock PR, et al: Oxygen radicals mediate endothelial cell damage by complement-stimulated granulocytes. J Clin Invest 1978;61:1161–1167.

199 Weiss SJ, Young J, LoBuglio AF, et al: Role of hydrogen peroxide in neutrophil-mediated destruction of cultured endothelial cells. J Clin Invest 1981;68:714–721.

200 Martin WJ II: Neutrophils kill pulmonary endothelial cells by a hydrogen-peroxide-dependent pathway. An in vitro model of neutrophil-mediated lung injury. Am Rev Respir Dis 1984;130:209–213.

201 Varani J, Fligiel SEG, Till GO, et al: Pulmonary endothelial cell killing by human neutrophils. Possible involvement of hydroxyl radical. Lab Invest 1985;53:656–663.

202 Shasby DM, Shasby SS, Peach MJ: Granulocytes and phorbol myristate acetate increase permeability to albumin of cultured endothelial monolayers and isolated perfused lungs. Role of oxygen radicals and granulocyte adherence. Am Rev Respir Dis 1983;127:72–76.

203 Harlin JM, Schwartz BR, Reidy MA, et al: Activated neutrophils disrupt endothelial monolayer integrity by an oxygen radical-independent mechanism. Lab Invest 1985;52:141–150.

204 Sugahara K, Cott GR, Parsons PE, et al: Epithelial permeability produced by phagocytosing neutrophils in vitro. Am Rev Respir Dis 1986;133:875–881.

205 Ayars GH, Altman LC, Rosen H, et al: The injurious effect of neutrophils on pneumocytes in vitro. Am Rev Respir Dis 1984;130:964–973.

206 Shasby DM, Lind SE, Shasby SS, et al: Reversible oxidant-induced increases in albumin transfer across cultured endothelium: Alterations in cell shape and calcium homeostasis. Blood 1985;65:605–614.

207 Spragg RG, Hinshaw DB, Hyslop PA, et al: Alterations in adenosine triphosphate and energy charge in cultured endothelial and $P388D_1$ cells after oxidant injury. J Clin Invest 1985;76:1471–1476.

208 Harlan JM, Callahan KS: Role of hydrogen peroxide in the neutrophil-mediated release of prostacyclin from cultured endothelial cells. J Clin Invest 1984;74:442–448.

209 Levine PH, Weinger RS, Simon JA, et al: Leukocyte-platelet interaction. Release of hydrogen peroxide by granulocytes as a modulator of platelet reactions. J Clin Invest 1976;57:955–963.

210 Schraufstätter IU, Hyslop PA, Hinshaw DB, et al: Hydrogen peroxide-induced injury of cells and its prevention by inhibitors of poly(ADP-ribose) polymerase. Proc Natl Acad Sci USA 1986;83:4908–4912.

211 Ager A, Gordon JL: Differential effects of hydrogen peroxide on indices of endothelial cell function. J Exp Med 1984;159:592–603.

212 Bates EJ, Johnson CC, Lowther DA: Inhibition of proteoglycan synthesis by hydrogen peroxide in cultured bovine articular cartilage. Biochim Biophys Acta 1985;838:221–228.

213 Bates EJ, Lowther DA, Johnson CC: Hyaluronic acid synthesis in articular cartilage: An inhibition by hydrogen peroxide. Biochem Biophys Res Commun 1985;132:714–720.

214 Lands WEM: Interactions of lipid hydroperoxides with eicosanoid biosynthesis. J Free Radicals Biol Med 1985;1:97–102.

215 Nathan CF, Arrick BA, Murray HW, et al: Tumor cell antioxidant defenses. Inhibition of the glutathione redox cycle enhances macrophage-mediated cytolysis. J Exp Med 1981;153:766–782.

216 Arrick BA, Nathan CF, Griffith OW, et al: Glutathione depletion sensitizes tumor cells to oxidative cytolysis. J Biol Chem 1982;257:1231–1237.

217 Harlan JM, Levine JD, Callahan KS, et al: Glutathione redox cycle protects cultured endothelial cells against lysis by extracellularly generated hydrogen peroxide. J Clin Invest 1984;73:706–713.

218 O'Donnell-Tormey J, DeBoer CJ, Nathan CF: Resistance of human tumor cells in vitro to oxidative cytolysis. J Clin Invest 1985;76:80–86.

219 Agar NS, Sadrzadeh SMH, Hallaway PE, et al: Erythrocyte catalase. A somatic oxidant defense. J Clin Invest 1986;77:319–321.

220 Winterbourn CC, Stern A: Human red cells scavenge extracellular hydrogen peroxide and inhibit formation of hypochlorous acid and hydroxyl radical. J Clin Invest 1987;80:1486–1491.

221 Van Asbeck B, Hoidal S, Vercellotti GM, Protection against lethal hyperoxia by tracheal insufflation of erythrocytes: role of cell glutathione. Science 1985;227:756–759.

222 Toth KM, Clifford DP, Berger EM, et al: Intact human erythrocytes prevent hydrogen peroxide-mediated damage to isolated perfused rat lungs and cultured bovine pulmonary artery endothelial cells. J Clin Invest 1984;74:292–295.

223 Havemann K, Janoff A (eds): Neutral Proteases of Human Polymorphonuclear Leukocytes. Munich, Urban & Schwarzenburg, 1978.

224 Vissers MCM, Winterbourn CC: Contribution of oxidants to neutrophil-mediated degradation of basement membrane collagen. Biochim Biophys Acta 1986;889:277–286.

225 McCord JM: Free radicals and inflammation: Protection of synovial fluid by superoxide dismutase. Science 1974;185:529–530.

226 Greenwald RA, Moy WW: Effect of oxygen-derived free radicals on hyaluronic acid. Arthritis Rheum 1980;23:455–463.

227 Wong SF, Halliwell B, Richmond R, et al: The role of superoxide and hydroxyl radicals in the degradation of hyaluronic acid induced by metal ions and by ascorbic acid. J Inorg Biochem 1981;14:127–134.

228 Carlin G, Djursäter R: Xanthine oxidase induced depolymerization of hyaluronic acid in the presence of ferritin. FEBS Lett 1987;177:27–30.

229 Greenwald RA, Moy WW: Inhibition of collagen gelation by action of the superoxide radical. Arthritis Rheum 1979;22:251–259.

230 Curran SF, Amoruso MA, Goldstein BD, et al: Degradation of soluble collagen by ozone or hydroxyl radicals. FEBS Lett 1984;176:155–160.

231 Dean RT, Roberts CR, Forni LG: Oxygen-centred free radicals can efficiently degrade the polypeptide of proteoglycans in whole cartilage. Biosci Rep 1984;4:1017–1026.

232 Bates EJ, Harper GS, Lowther DA, et al: Effect of oxygen-derived reactive species on cartilage proteoglycan-hyaluronate aggregates. Biochem Int 1984;8:629–637.

233 Wolff SP, Dean RT: Fragmentation of proteins by free radicals and its effect on their susceptibility to enzymic hydrolysis. Biochem J 1986;234:399–403.

234 Weiss SJ, Regiani S: Neutrophils degrade subendothelial matrices in the presence of alpha-1-proteinase inhibitor: cooperative use of lysosomal proteinases and oxygen metabolites. J Clin Invest 1984;73:1297–1303.

235 Baehner RL, Boxer LA, Allen JM, et al: Autooxidation as a basis for altered function by polymorphonuclear leukocytes. Blood 1977;50:327–335.

236 Roos D, Weening RS, Voetman AA, et al: Protection of phagocytic leukocytes by endogenous glutathione: Studies in a family with glutathione reductase deficiency. Blood 1979;53:851–866.

237 Clark RA: Modulation of the inflammatory response by the neutrophil myeloperoxidase system. Adv Exp Med Biol 1982;141:207–216.

238 Dri P, Soranzo MR, Cramer R, et al: Role of myeloperoxidase in respiratory burst of human polymorphonuclear leukocytes. Studies with myeloperoxidase-deficient subjects. Inflammation 1985;9:21–31.

239 Jandl RC, Andre-Schwartz J, Borges-DuBois L, et al: Termination of the respiratory burst in human neutrophils. J Clin Invest 1978;61:1176–1185.

240 Edwards SW, Swan TF: Regulation of superoxide generation by myeloperoxidase during the respiratory burst of human neutrophils. Biochem J 1986;237:601–604.

241 Stendahl O, Coble B-I, Dahlgren C, et al: Myeloperoxidase modulates the phagocytic activity of polymorphonuclear neutrophil leukocytes. Studies with cells from a myeloperoxidase-deficient patient. J Clin Invest 1984;73:366–373.

242 Lee CW, Lewis RA, Corey EJ, et al: Oxidative inactivation of leukotriene C_4 by stimulated human polymorphonuclear leukocytes. Proc Natl Acad Sci USA 1982;79:4166–4170.

243 Coble B-I, Dahlgren C, Hed J, et al: Myeloperoxidase reduces the opsonizing activity of immunoglobulin G and complement component C3b. Biochim Biophys Acta 1984;802:501–505.

244 Voetman AA, Weening RS, Hamers MN, et al: Phagocytosing human neutrophils inactivate their own granular enzymes. J Clin Invest 1981;67:1541–1549.

245 Clark RA, Borregaard N: Neutrophils autoinactivate secretory products by myeloperoxidase-catalyzed oxidation. Blood 1985;65:375–381.

246 Vissers MCM, Winterbourn CC: Myeloperoxidase-dependent oxidative inactivation of neutrophil proteinases and microbicidal enzymes. Biochem J 1987;245:277–280.

247 Weiss SJ, Peppin G, Ortiz X, et al: Oxidative autoactivation of latent collagenase by human neutrophils. Science 1985;227:747–749.

248 Peppin GJ, Weiss SJ: Activation of the endogenous metalloproteinase, gelatinase, by triggered human neutrophils. Proc Natl Acad Sci USA 1986;83:4322–4326.

249 Shah SV, Baricos WH, Basci A: Degradation of human glomerular basement membrane by stimulated neutrophils. Activation of a metalloproteinase(s) by reactive oxygen metabolites. J Clin Invest 1987;79:25–31.

250 Vissers MCM, Winterbourn CC: Activation of human/endogenous neutrophil gelatinase by co-released serine proteinases. 1987;249:325–329.

251 Agner K: Detoxicating effects of verdoperoxidase on toxins. Nature 1947;159:271–272.

252 Clark RA: Oxidative inactivation of pneumolysin by the myeloperoxidase system and stimulated human neutrophils. J Immunol 1986;136:4617–4622.

253 Clark RA, Leidal KG, Taichman NS: Oxidative inactivation of *Actinobacillus actinomycetemcomitans* leukotoxin by the neutrophil myeloperoxidase system. Infect Immun 1986;53:252–256.

254 El-Hag A, Lipsky PE, Bennett M, et al: Immunomodulation by neutrophil myeloperoxidase and hydrogen peroxide: Differential susceptibility of human lymphocyte functions. J Immunol 1986;136:3420–3426.

255 Dallegri F, Patrone F, Frumento G, et al: Down-regulation of K cell activity by neutrophils. Blood 1985;65:571–577.

256 Seaman WE, Gindhart TD, Blackman MA, et al: Suppression of natural killing in vitro by monocytes and polymorphonuclear leukocytes. Requirement for reactive metabolites of oxygen. J Clin Invest 1982;69:876–888.

257 Carp H, Janoff A: Potential mediators of inflammation. Phagocyte-derived oxidants suppress the elastase-inhibitory capacity of alpha-1-proteinase inhibitor in vitro. J Clin Invest 1980;66:987–995.

258 Johnson D, Travis J: The oxidative inactivation of human α-1-proteinase inhibitor. Further evidence for methionine at the reactive center. J Biol Chem 1979;254:4022–4026.

259 George PM, Vissers MCM, Travis J, et al: A genetically engineered mutant of α_1-antitrypsin protects connective tissue from neutrophil damage and may be useful in lung disease. Lancet 1984;ii:1426–1428.

C.C. Winterbourn, PhD, Department of Pathology, Christchurch School of Medicine, Christchurch Hospital, Christchurch (New Zealand)

3 Oxidant Radical Production and Lung Injury

Henry Jay Forman

During the past 15 years, free radicals have been implicated increasingly in the mechanism of toxicity of a variety of agents including paraquat, hyperoxia, carbon tetrachloride, NO_2, O_3, and anticancer drugs, such as bleomycin and adriamycin. While some of these agents can cause direct oxidation of cellular components, others may cause an increase in the generation of oxidant species, which are more reactive than the original agent itself. Among such agents are paraquat and hyperoxia, both of which are believed to act as poisons through the intracellular generation of oxygen-centered free radicals. This chapter, which concerns mechanisms of oxidant radical production and lung injury, will focus upon paraquat and hyperoxia since both of these agents are associated primarily with pulmonary pathophysiology.

In this chapter, I have described briefly the pathophysiology of hyperoxia and paraquat toxicity, the general theory of free radical toxicity, and antioxidant defenses. This is followed by a review of recent studies from my laboratory, which relate to these topics.

Pathophysiology

When pure O_2 (at one atmosphere absolute pressure) is breathed for prolonged periods, death as a consequence of alveolar flooding is the usual consequence [1, 2]. For rats, this exposure results in death from pulmonary edema of 50% of the animals in 66–72 h [2]. The lung is the only internal organ which is in direct contact with the atmosphere. For other organs, the principal source of tissue O_2 is that carried by hemoglobin. Hemoglobin is nearly saturated with O_2 even under normoxic condition. In hyperoxia, the increase in O_2 that reaches internal organs, other than the lung, is therefore due to only the increased O_2 content in the plasma. As the solubility of O_2 in plasma is quite low, the actual amount of O_2 reaching these organs in hyperoxia is low in comparison with that reaching the lung. Although the greater O_2 concentration reaching the lung may account, in part, for the relative susceptibility of the lung to normobaric O_2 toxicity, other organs can be adversely affected by hyperoxia [3]. Above 2.5 atmospheres absolute

O_2 approximately, central nervous system toxicity becomes the predominant problem and death may result from seizures. In this chapter we will, however, focus upon normobaric O_2 toxicity and the consequences for lung cells.

Morphological and functional changes in lungs develop between 24 and 48 h of O_2 exposure [4]. Distortion of the plasma and mitochondrial membranes in the endothelium appear to be the first cellular change [5–7]. Alveolar macrophages are also damaged early as indicated by functional losses [8–11] (see Alveolar Macrophage Studies). Subsequently, type I epithelial cells are damaged and lost from the epithelial surface. The type I epithelial cell is the very thin cell that constitutes most of the air side of the air/blood interface in the lung. In contrast, the granular (type II) pneumocyte is relatively resistant to hyperoxic injury and in sublethal oxygen toxicity proliferates [12, 13] and repopulates the alveolar epithelium [12]. In the normal lung parenchyma the diffusion barrier for gas exchange is negligible. The type II cell is cuboidal and, thus, replacement of type I by type II cells presents a significantly increased diffusion barrier in the alveolus. In the latter stages of normobaric O_2 toxicity, neutrophils and/or circulating monocytes appear in increasing numbers in the lung [14, 15]. Although it has been suggested that neutrophils could be involved in lung damage during hyperoxic exposure, the evidence suggests that the neutrophils do not participate in the ultimate pathology from normobaric oxygen toxicity [16]. Nevertheless, neutrophils may participate in the development of fibrosis or the restructuring of the lung that occurs from exposure to sublethal hyperoxia.

The production of of O_2^- and H_2O_2 by the alveolar macrophage and polymorphonuclear leukocyte has been implicated in tissue damage in inflammation. Although the metabolic and functional alterations occurring in such externally applied oxidant stress, may be similar to that occurring in hyperoxia, the primary focus of this chapter is on the effects of intracellular generation of free radicals.

The pathophysiology of paraquat toxicity is somewhat different from that of hyperoxia. At a dose of paraquat where the plasma levels are in the millimolar range all organs may be susceptible to paraquat toxicity and animals appear to die primarily from acute renal failure [17]. At lower doses or with survival from the acute toxicity, paraquat acts as a slow poison with death resulting from pulmonary edema or necrosis days or weeks after the initial insult. Paraquat and hyperoxia are alike in that the primary target of chronic toxicity is the lung, but the two agents differ in the principal cellular target within the lung. While the type II cell is relatively resistant in hyperoxia, this cell appears to be a specific target in paraquat toxicity [18]. Paraquat toxicity therefore diminishes two vital maintenance functions of the lung since type II cells are both the producer of surfactant and the precursor of the type I epithelial cell. With sublethal paraquat toxicity, fibrosis can also occur [19].

The Free Radical Theory of Oxidant Injury

There have been several recent comprehensive reviews concerning the role of free radicals and antioxidant defenses in lung injury [20–25]. From the accumulated evidence, it is now clear that reduced forms of O_2, rather than O_2 itself are the toxic agents in O_2 toxicity [26–28]. Superoxide anion (O_2^-) is produced during normal cellular metabolism [26, 27]. H_2O_2 is produced in several enzyme catalyzed reactions [29, 30] and through the dismutation of O_2^- so that any circumstance resulting in increased O_2^- production would also result in increased H_2O_2 output (fig. 1).

In some biological reactions the rate of O_2^- generation increases as a function of O_2 concentration [31–37]. Normally, in aerobic cells, over 95% of O_2 consumption occurs in the mitochondrial production of adenosine triphosphate (ATP). In normoxia, only a small fraction of the electrons that enter the ATP-generating mitochondrial respiratory chain escape before being transferred to cytochrome oxidase. This 'terminal oxidase' then transfers all the electrons to O_2 without formation of any intermediate reduced O_2 byproducts. The small amount of electrons that escape from the respiratory chain do so by nonenzymatically catalyzed reaction with O_2 and account for much of the cellular O_2^- generation that occurs in normoxia [reviewed in ref. 36]. There are two major sites of O_2^- production in the respiratory chain [36]: NADH dehydrogenase and the ubiquinone-cytochrome b-cytochrome c_1 complex. Under normoxia the leak of electrons resulting in the generation of O_2^- is small. The reaction of O_2 with ubisemiquinone (QH$^{\cdot}$), a normal intermediate in the respiratory chain is a thermodynamically unfavorable reaction:

$$QH^{\cdot} + O_2 + H^+ \xleftarrow{\hspace{1.5cm}} QH_2 + O_2^-$$

Nevertheless, this reaction can be 'pulled forward' by removal of the product, O_2^-, by either the action of superoxide dismutase:

$$2\,O_2^- + 2\,H^+ \dashrightarrow H_2O_2 + O_2$$

or by reaction with targets or scavenging antioxidants, such as ascorbate (see Antioxidant Defenses). In hyperoxia, however, the reaction of ubisemiquinone with O_2 preceeds at an increased rate thereby increasing

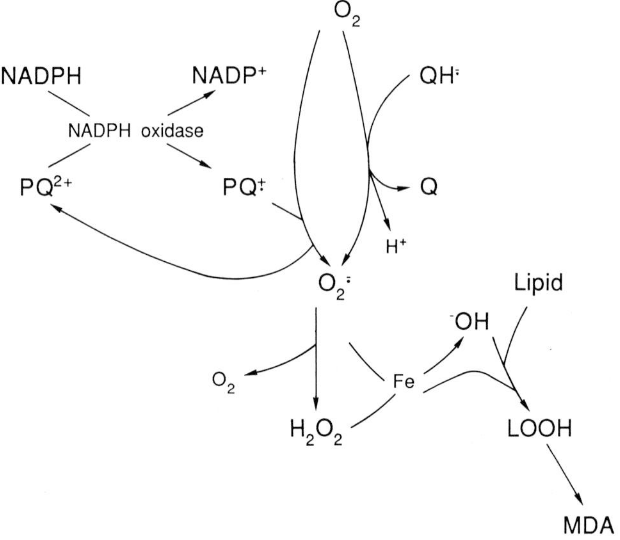

a

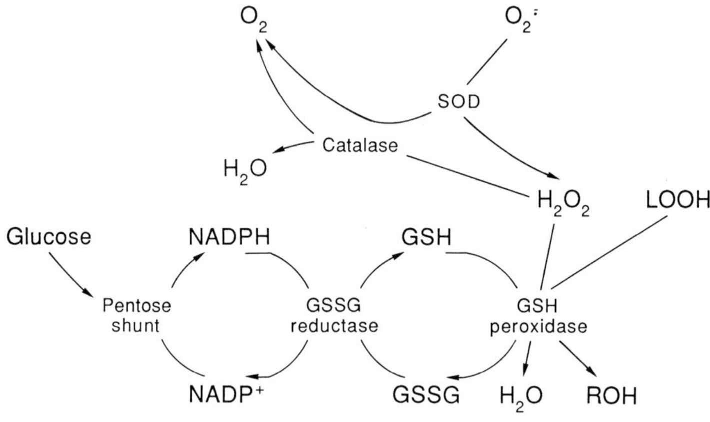

b

Fig. 1. a The generation of free radicals from oxygen or paraquat. Oxygen can be reduced in the mitochondrial respiratory chain by reaction with ubisemiquinone. Paraquat is reduced by an NADPH oxidase to paraquat radical. The paraquat radical then can react with oxygen to generate superoxide. Superoxide reacts with itself in a dismutation reaction. This reaction produces hydrogen peroxide and oxygen. Superoxide plus hydrogen peroxide and iron react to produce hydroxyl radical. Hydroxyl radical or iron-oxygen complexes can react with lipid to produce the lipid hydroperoxides which are responsible for membrane damage. Lipid hydroperoxides can break down to produce a number of compounds including malondialdyde (MDA).

the steady state O_2^- concentration and this accounts for most of the increase in O_2^- observed in hyperoxic exposure of mitochondria [33]. Freeman and co-workers [34, 35] have shown that cyanide insensitive respiration, a semiquantitative measure of O_2^- and H_2O_2 production increases in lung homogenates and lung mitochondria in response to hyperoxic exposure in the range where toxic effects on the whole animal are observed. Thus, the normally thermodynamically unfavorable production of O_2^- can be kinetically 'pushed forward' in hyperoxia.

Paraquat is reduced in cells in an NADPH-dependent enzyme-catalyzed process [38, 39]. Interestingly, the reduction of paraquat (PQ^{2+}) by NADPH:

$$2\,PQ^{2+} + NADPH \;\rightleftharpoons\; NADP^+ + 2\,PQ^+ + H^+$$

is also a thermodynamically unfavorable reaction that is 'pulled forward' by removal of the product. In this case, the product, the paraquat radical (PQ^+), reacts with molecular O_2 in an extremly rapid nonenzymatic reaction which generates O_2^- [40]:

$$PQ^- + O_2 \;-----\rightarrow\; PQ^{2+} + O_2^-$$

The rate-limiting step in the generation of O_2^- in paraquat toxicity is the reduction of paraquat in the enzymatic process. In cases where simultaneous paraquat and hyperoxia cause a synergistic toxicity, it is therefore unlikely that the increase in O_2 caused a greater increase in paraquat-mediated O_2^- production. Rather, it is likely that the combination of epithelial and endothelial injuries intensifies the pathological consequences under these conditions (see Paraquat Studies).

According to the free radical theory of oxidant injury, which includes H_2O_2 as an honorary member of the radical establishment, the production of O_2^- and H_2O_2, if unchecked by antioxidant defenses eventually leads to oxidation of cellular constituents and destruction of cell function. Initiation

---

Fig. 1. b The reactions of the antioxidant enzymes. Superoxide dismutase (SOD) can catalyze the reaction of superoxide to produce hydrogen peroxide and oxygen at a rate which is 10^4 times that of the noncatalyzed reaction. Catalase can catalyze the reaction of hydrogen peroxide to form water and oxygen. Glutathione peroxidase can remove hydrogen peroxide or lipid hydroperoxide by reducing them with glutathione to produce water or the lipid alcohal respectively. The oxidized glutathione (GSSG) can then be re-reduced by the action of glutathione reductase using NADPH as a source of reducing equivalents. NADPH is then regenerated with pentose shunt enzymes which are dependent ultimately on glucose as a substrate.

of lipid peroxidation, which may involve the formation of hydroperoxyl radical ($HO_2^{\cdot}$, protonated superoxide) or hydroxyl radical ($^{\cdot}OH$) or iron-O_2 complexes [41–46], may be the critical event in oxidant injury. Lipid peroxidation poses a threat to the integrity of membranes, metabolic integration and the antioxidant system itself (see Antioxidant Defenses). In addition, recent evidence suggests that cyclooxygenase and lipoxygenase activities are also stimulated by peroxides [47, 48]. The enzymatically catalyzed lipid peroxidation by these two enzymes results in the formation of prostaglandins, leukotrienes and related compounds. These products are extremely effective agonists for a variety of physiologic actions, including changes in vascular permeability, vasodilation and/or vasoconstriction, and chemotaxis for phagocytes. Thus the dangers from lipid peroxidation induced by free radical production may be magnified by production of prostaglandins and leukotrienes. Perhaps it is not too radical to propose that pulmonary edema is mediated by production of enzymatically generated factors rather than by wholesale disruption of the epithelial barrier through non-enzymatic lipid peroxidation.

One of the first proposals for the toxic effects of O_2 was inactivation of enzymes [49]. Attention has now been refocused on this possibility since oxidation of amino acid moieties may be a part of the normal turnover process of enzymes [50]. It has been recognized for some time that the Cu-containing superoxide dismutase is inactivated by oxidation of active site histidines [51] and that H_2O_2 inactivates superoxide dismutase by such a mechanism [52, 53]. More recently, it has been shown that O_2^{-} inactivates catalase [54]. It has been suggested that such inactivation may account for the decrease in superoxide dismutase and catalase activity that occurs during prolonged O_2 exposure of the guinea pig alveolar macrophage [55].

Cellular effects of prolonged hyperoxia include: the inactivation of reverse electron transport through the mitochondrial respiratory chain, which may be important in regulating the cellular $NADH/NAD^+$ ratio [56], decreased Na-K-ATPase activity [57, 58], decreased ATP production [59, 60], decreased protein synthesis [4], and decreased uptake of serotonin by the lung endothelium [61, 62]. Peroxides lower the $NADPH/NADP^+$ ratio and alter cellular distribution of glutathione such that the reduced to oxidized glutathione ratio is lowered while glutathione-protein conjugate formation increases [63–66]. These alterations in the cellular redox state can cause disruption of Ca^{2+} distribution between organelles and the cytosol [63–66]; however, the mechanism of these effects are still uncertain [67, 68]. Ca^{2+} distribution between various intracellular compartments and the extracellular medium is also affected by alterations of membrane potentials and/or ATP concentration [69, 70]. As cellular distribution of Ca^{2+} and the cellular redox state are involved in regulation of numerous aspects of metabolism and function, their alteration by peroxides has obvious implications in the mechanism of oxidant injury. For example, a transient increase in intracellular Ca^{2+} can trigger arachidonate release and metabolism, and increase activities of protein kinases, lipases, and phosphatases. Thus, most of the metabolic functions that oxidant stress may alter are interdependent and therefore affect the regulation of metabolism and cellular function. What has not been established is the relative timing of these alterations so that cause and effect relationships can be determined.

Antioxidant Defenses

Figure 1 shows a general scheme for antioxidant defenses and related metabolic consequences of oxidant stress. Enzyme catalyzed elimination of O_2^- by superoxide dismutase and H_2O_2 by catalase or glutathione peroxidase is probably the major line of antioxidant defense. The removal of O_2^- and H_2O_2 by the reactions:

$$2\,H_2O_2 \xrightarrow{\text{catalase}} 2\,H_2O + O_2$$

$$H_2O_2 + GSH \xrightarrow{\text{glutathione peroxidase}} H_2O + GSSG$$

where GSH is reduced glutathione and GSSG is oxidized glutathione. Once formed, lipid peroxides can oxidize glutathione through the action of glutathione peroxidase:

$$ROOH + GSH \xrightarrow{\text{glutathione peroxidase}} ROH + GSSG$$

where R is a lipid moiety. Reduced glutathione is regenerated from the oxidized form by glutathione reductase with NADPH:

$$NADPH + GSSG + H^+ \xrightarrow{\text{glutathione reductase}} 2\,GSH + NADP^+$$

The glutathione system may also restore other oxidized tissue components to their reduced state [71]. NADPH levels are maintained by pentose phosphate shunt activity. The supply of glutathione and NADPH may be of critical importance for antioxidant defense. With oxidant stress the redox state of the cell could become oxidized and the cell would then have inadequate antioxidant defenses to prevent irreversible damage, such as lipid peroxidation. Once oxidant stress is stopped, however, the redox state itself may return to normal while other cellular components remain irreversibly altered (see Alveolar Macrophage Studies).

There have been numerous studies indicating a positive correlation of increased antioxidant enzymes in lungs and lung cells with increased survival in hyperoxia [5, 72–90] and/or that decreased antioxidant defenses correlate with decreased survival [91–95]. Constitutive levels of several antioxidant enzymes are relatively higher in type II cells than in the rest of the lung and this may also partially account for the relative resistance of the type II cell to hyperoxia [75]. Antioxidant enzymes can be increased in the lung or endothelial cells in culture using liposomes and that such administration has protective effects [88, 89].

A useful model for studying antioxidant defense systems is adaptation to hyperoxia. When adult rats are exposed to sublethal hyperoxia in the range of 80–90% O_2 at one atmosphere for 5 or more days, alterations in lung structure [96] and increases in lung antioxidant enzymes occur [5, 72–75, 79–81, 83, 86, 87, 89] along with a marked increase in tolerance to subsequent exposure to 100% O_2. An even more rapid adaptation to hyperoxia can be produced in the rat by administration of endotoxin along with exposure to 100% O_2 [82, 84].

Since adaptation to hyperoxia has been correlated with increases in antioxidant enzymes including the selenium-containing enzyme, glutathione peroxidase, a study was made to determine whether selenium deficiency would prevent an increase in glutathione peroxidase during O_2 exposure and thereby inhibit adaptation [93]. We also manipulated the sulfur amino acid content of the diet. We assume that our prolonged cysteine and methionine restricted diet resulted in a similar effect to that observed by Deneke et al. [95]. These investigators have shown that short term cysteine deficiency caused a failure of the lungs of rats exposed to hyperoxia to increase in glutathione, an event which usually occurs during adaptation [79]. Figure 2 demonstrates the mortality of the rats in the various dietary groups in 80% O_2, a dose of O_2 that rats generally survive and which produces adaptation to 1 atmosphere absolute O_2. Without adaptation, the LD_{50} for 1 atmosphere absolute O_2 is 72 h [2]. While both selenium and sulfur-containing amino acid deficiency led to increased mortality in 0.8 atmosphere absolute O_2, all but one of the surviving rats survived subsequent 96-hour exposure to 1 atmosphere absolute O_2.

Glutathione peroxidase activity was decreased by the low selenium diet (fig. 3). After O_2 exposure, lung glutathione peroxidase activity was elevated in all dietary groups with the largest percentage increase being in the low selenium rats. Nonetheless, even after exposure to O_2, the resulting glutathione peroxidase activity in the low selenium rats was only 34–70% of that in the unexposed high selenium rats. The absolute content of

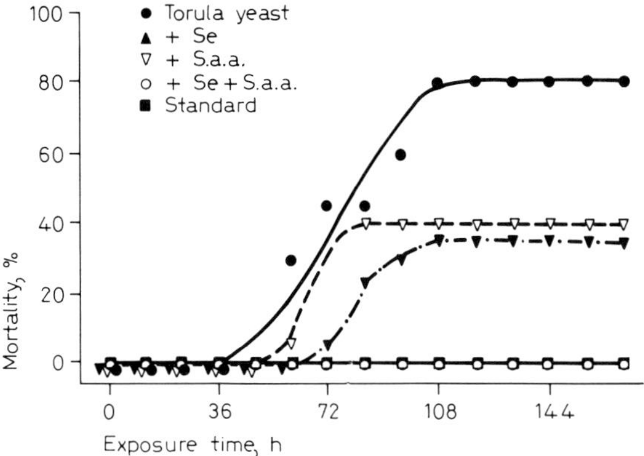

Fig. 2. Effect of exposure to 80% oxygen on rats fed Torula yeast diets with or without supplementation with either sodium selenite (Se) and/or sulfur-containing amino acids (S.a.a.) is shown. There were 20 rats in each of the low S.a.a. groups and 10 rats in each of the high S.a.a. groups. Four rats, fed a standard laboratory diet, were included for comparison. From Forman et al. [93].

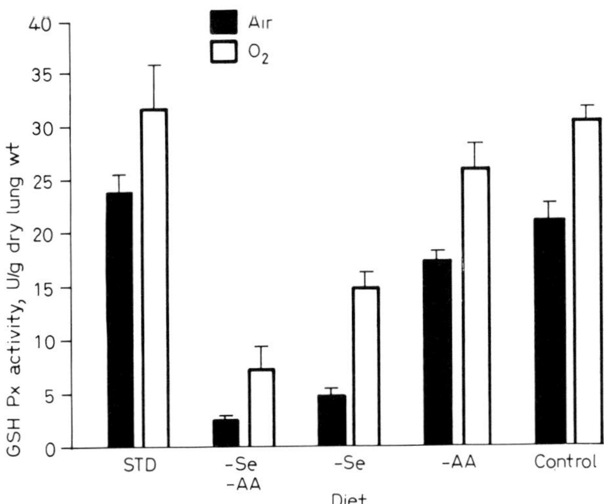

Fig. 3. Glutathione peroxidase activity in selenium and sulfur amino acid-deficient diets before and after oxygen exposure. The activities shown are for whole lung from either control rats or from rats exposed to 80% oxygen for 7 days. The activities are from survivors only. AA indicates sulfur amino acid content of the diet. Controls were the Torula yeast diet with selenium and sulfur-containing amino acids added.

glutathione peroxidase activity therefore can be correlated with decreased resistance to hyperoxia but adaptation appears to correlate more with the increase in the enzyme than with the absolute activity. Thus this study demonstrates that both glutathione peroxidase and glutathione are important in the defense against O_2 toxicity.

Nonenzymatic defenses include agents such as vitamin E, ascorbate, glutathione and other sulfhydryl compounds, which can react with oxygen-derived radicals directly [see ref. 22 for review]. Recently, it has been shown that urate can prevent iron-catalyzed oxidation of ascorbate in blood without being oxidized itself [97], possibly, through chelation of iron (see above).

Paraquat Studies

What is the cause for difference in relative cellular susceptibility to paraquat and hyperoxia? Differences in antioxidant enzyme activities [75] probably contribute significantly to the relative sensitivity of lung cells to oxidant challenge. In this regard, it is puzzling that the type II cell, which has relatively high antioxidant enzyme activities and is relatively resistant to hyperoxia, is the major target for paraquat toxicity [18].

Possible causes for the differences in the susceptibility of lung cells to paraquat toxicity were investigated with isolated lung cells and perfused lungs [39]. Uptake of paraquat was measured as a function of both time and concentration (between 1×10^{-6} and $1 \times 10^{-3}\,M$). With alveolar macrophages, uptake of paraquat with time reached a plateau as internal approached external concentration. In alveolar macrophages this equilibration between internal and external paraquat concentration occurred at all initial concentrations. With granular pneumocytes, uptake was nonlinear with external concentration. In granular pneumocytes, accumulation above external concentration occurred; however, in the presence of inhibitors of ATP synthesis, intracellular paraquat concentration reached a plateau as it approached the extracellular paraquat concentration (fig. 4).

Apparently, paraquat can enter both cell types by non-energy-dependent facilitated transport, but granular pneumocytes can also accumulate paraquat by an energy-dependent process. With isolated perfused lungs, uptake was nonlinear with time; in the first hours a plateau occurred when the intracellular approached extracellular ($1 \times 10^{-5}\,M$) concentration, but a slower second process resulted in a 6-fold accumulation by 4 h.

The effect of paraquat on NADPH oxidation was measured with the supernatant fraction from cells and lungs. The maximal rate of paraquat-dependent NADPH oxidation was about 2 times higher in granular pneu-

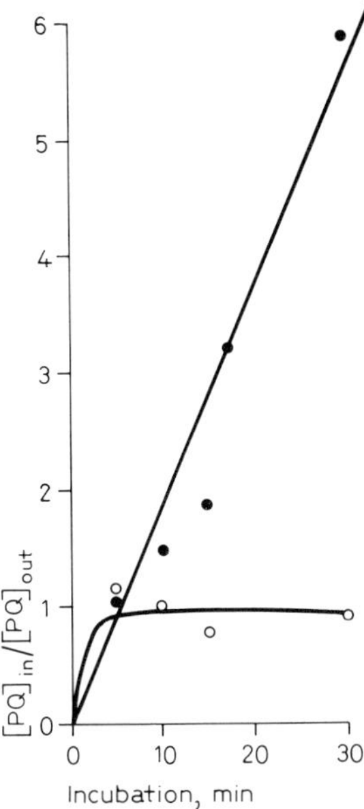

Fig. 4. The uptake of paraquat (PQ) by granular pneumocytes. The line with the solid dots indicates paraquat uptake by type II cells when extracellular paraquat is $4.5 \times 10^6\ M$. The curve with the open circle represents paraquat uptake in the presence of 5 mM glucose and antimycin A at 400 ng/ml. From Forman et al. [39].

mocytes than in alveolar macrophages or whole lungs, but the estimated enzyme kinetic constants were similar in the two cell homogenates (K_m for paraquat was $\sim 2 \times 10^{-4}\ M$ and K_m for NADPH was $\sim 5 \times 10^{-6}\ M$. The results suggest the paraquat-dependent NADPH oxidation rates could differ markedly between intact cells depending upon the extracelluar paraquat concentration. At low paraquat concentration, energy-dependent transport into granular pneumocytes would favor a significantly higher paraquat concentration in those cells while alveolar macrophages would have only the low paraquat concentration. At high extracellular paraquat concentrations the intracellular paraquat concentrations would be fairly close in the two cell types. Therefore at very high paraquat levels, which may be observed in acute paraquat toxicity, the rates of paraquat-induced NADPH oxidation and O_2^- production could be close in most cell types. At the low extracellular paraquat concentrations that are observed in chronic paraquat toxicity, only those cells which actively accumulate this agent would have sufficient paraquat concentration for the toxic processes to occur.

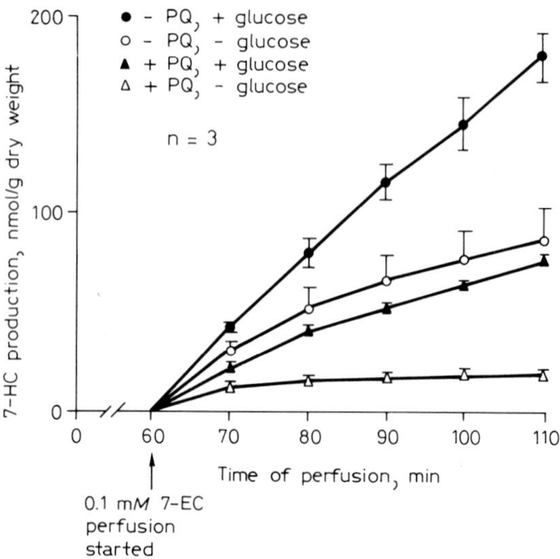

Fig. 5. 7-Hydroxycoumarin (7-HC) production (after 60-min equilibration) in lungs perfused with 0.1 mM 7-ethoxycoumarin (7-EC) and additions as noted in the key. When present, concentrations were glucose 5 mM and paraquat (PQ) 1 mM. Paraquat inhibited 7-HC production in either the presence or absence of glucose.

In another investigation [98], we asked if NADPH depletion could play a role in the toxicity of paraquat. The effect of paraquat on 7-ethoxycoumarin deethylation, an NADPH-dependent cytochrome P-450-linked reaction, was studied in isolated perfused rat lungs. When perfused with glucose (control), lungs oxidatively deethylated 7-ethoxycoumarin at a constant rate. When glucose was omitted from the perfusate, 7-ethoxycoumarin deethylation was significantly lower and the rate progressively decreased, suggesting inadequate pentose shunt regeneration of NADPH (fig. 5). During perfusion with paraquat ($10^{-3}\,M$), 7-ethoxycoumarin metabolism was still lower. Paraquat concentration (between 10^{-6} and $10^{-3}\,M$) in the perfusate inhibited 7-ethoxycoumarin deethylation in a dose-dependent fashion but did not inhibit the enzyme in isolated microsomes (measured in the presence of an NADPH-regenerating system). Thus, in the intact isolated lung, paraquat depletes NADPH to a sufficient extent to impair mixed-function oxidation.

Paraquat has been reported to initiate lipid peroxidation in isolated perfused lungs [99]. If lipid hydroperoxides are not rapidly removed, a breakdown product, malondialdehyde will be produced and will efflux from cells (fig. 1). Efflux of oxidized glutathione during oxidant stress is an expected consequence of increased turnover of glutathione peroxidase since the steady state level of oxidized glutathione will be increased. These effects can be observed by monitoring the content of malondialdehyde and oxidized glutathione in the perfusate of isolated perfused lungs. In normal lungs perfused with paraquat, there was a marked efflux of oxidized glutathione (83 nmol in 90 min) compared with lungs from selenium-deficient rats (14 nmol in 90 min) or nonparaquat exposed lungs (30 nmol in 90 min). Selenium deficiency, which decreased glutathione peroxidase activity 75–80%, caused a marked decrease in glutathione oxidation because H_2O_2 and/or lipid peroxides produced as a result of paraquat redox cycling were not removed at the expense of reduced glutathione. The efflux of malondialdehyde from the selenium-deficient, paraquat-treated rat lungs (11 nmol in 90 min) was also significantly greater than the efflux from other groups (approximately 0). Thus, in lungs with normal glutathione peroxidase activity, paraquat-induced malondialdehyde production was prevented by removal of lipid hydroperoxides; but, consequently, this results in the consequent generation of excess oxidized glutathione. In the selenium-deficient lungs, paraquat-induced lipid peroxidation resulted in malondialdehyde production and lowered oxidized glutathione efflux. These studies further support a role for the glutathione pathway in protection against oxidant stress.

Alveolar Macrophage Studies

The alveolar macrophage is an important component of the defense system of the lung against foreign organisms and tumor cells. Alveolar macrophages are endowed with the metabolic capacity to generate agents for destruction of microorganisms (e.g. oxygen radicals produced during the 'respiratory burst' [100] and for recruitment of neutrophils [101–103]. These processes can be stimulated by a wide variety of agents including zymosan, lectins, latex beads, A23187 (a Ca^{2+} ionophore), fluoride ion, and digitonin. While alveolar macrophages can be stimulated to released O_2^- into the extracellular fluid and O_2^- production is implicated in the mechanism of toxicities we have described, these processes are separable phenomena. Indeed, as will be discussed below, oxidant stress inhibits the ability of alveolar macrophages to produce O_2^-. In that respect, the respiratory burst is like other cellular metabolic functions, which can be adversely affected by oxidant stress.

Activation of the respiratory burst involves many of the processes that may be altered by hyperoxia. The respiratory burst begins through receptor binding or chemical perturbation of the plasma membrane [104]. The plasma membrane potential then depolarizes [105]. Changes in Ca^{2+}

distribution are also required for initiation as indicated by inhibition of O_2^- release by a calmodulin antagonist [106], by partial inhibition by Ca^{2+} channel blockers and by stimulation with A23187, a Ca^{2+} ionophore [107]. Ca^{2+} redistribution is involved in inositol polyphosphate and protein kinase C-mediated cell events [108, 109]. This process has been demonstrated in peritoneal macrophages from guinea pigs [110]. Recent evidence has suggested, however, that the rise in intracellular Ca^{2+} may be a concomitant but non-regulatory event [111]. More puzzlingly, free Ca^{2+} can also inhibit O_2^- production by phagocytic vesicles isolated from stimulated macrophages [106]. Activation of the oxidase responsible for O_2^- production also requires ATP [112]. The active oxidase requires NADPH for generation of O_2^- [113].

While it seems clear that alveolar macrophages are not the primary target for hyperoxic damage in the lung, a substantial number of studies have shown that alveolar macrophages from different species are damaged early in the course of in vivo or in vitro hyperoxic exposure as indicated by loss of respiratory burst activity, phagocytosis and chemotactic response (8–11, 114–117]. Human alveolar macrophages exposed to 0.95 atmosphere absolute O_2 in vitro show a progressive decrease in ATP content [118] after 48 h, but this was not correlated with any functional measurements. 18 h of hyperoxia has been demonstrated to increase the synthesis of fibronectin by human alveolar macrophages [119]. It has also been suggested that early damage to the alveolar macrophage may be responsible for the later influx of neutrophils [15]. These metabolic alterations may have significance in the chronic phase of O_2 toxicity. The focus of the studies that are described below is the effects of oxidants on the alveolar macrophage itself.

The alveolar macrophage is not a primary target for paraquat toxicity; nonetheless, high concentrations (1 mM in the following example) of this herbicide can alter alveolar macrophage metabolism and function (see below). Alveolar macrophages incubated with paraquat were inhibited in their ability to release superoxide when stimulated with substances which provoke the respiratory burst [113]. The O_2^- production capacity of cells incubated with paraquat was approximately 10% of that of cells incubated with glucose. Addition of glucose after incubation with paraquat completely reversed the loss of respiratory burst activity. As previous studies had indicated that the reduction of paraquat by lung microsomes was NADPH-dependent [120], we hypothesized that the inhibition of the respiratory burst by paraquat was due to a specific depletion of NADPH. This hypothesis was supported by evidence indicating that: (a) paraquat was reducible by NADPH but not by NADH; (b) NADPH oxidation is saturable with paraquat, but NADH oxidation was unaffected by paraquat;

(c) paraquat enhanced pentose phosphate shunt activity 51%; (d) mitochondrial respiration and the effect of glucose on respiration were unaffected by paraquat, and (e) ATP concentration in alveolar macrophages was unaffected by paraquat. When cells were incubated in the presence of paraquat but in the absence of glucose, the significantly lower superoxide production upon the addition of concanavalin A was therefore due to the specific depletion of NADPH rather than to any effect on NADH or ATP. The reversal by glucose of the paraquat effect involves an increase in pentose shunt activity to maintain or restore the NADPH concentration.

The relationship between the redox state of NADPH- and concanavalin A-stimulated superoxide production in rat alveolar macrophages has also been examined under conditions of oxidant stress brought about by exposure to nonlethal time and dose exposures to t-butyl hydroperoxide, a model for the effects of lipid hydroperoxide [121]. Alveolar macrophages were exposed for 15 min to 10 μM t-butyl hydroperoxide in the presence or absence of exogenous glucose (5 mM) after which cells were assayed for concanvalin A-stimulated O_2^- production or for NADPH and NADP$^+$ and the distribution of cellular glutathione. Alveolar macrophages incubated in the absence of glucose showed a decreased respiratory burst when exposed to t-butyl hydroperoxide as was observed with paraquat (see above) (fig. 6). In the presence of glucose, 10 μM t-butyl hydroperoxide did not inhibit concanavalin A-stimulated O_2^- production.

Cells incubated without glucose showed a relatively high percent oxidation state of the NADPH/NADP$^+$ couple ($\sim$58% NADP$^+$ versus 27% in presence of glucose) which rose to almost 100% NADP$^+$ following exposure to 10 μM t-butyl hydroperoxide. Cells incubated without glucose and exposed to t-butyl hydroperoxide showed an increase in glutathione-mixed disulfides, a dramatic change in the redox state of glutathione, and an increased loss of oxidized glutathione from the cells (table 1). In another series of experiments, glucose was added 14 min after t-butyl hydroperoxide. Extracellular glutathione was higher than for control (minus t-butyl hydroperoxide) or when glucose was present during incubation with t-butyl hydroperoxide but free intracellular glutathione was restored to control level. Addition of glucose after incubation with t-butyl hydroperoxide also restored the respiratory burst and the NADPH redox state. Thus metabolic and functional alterations due to transient oxidation of the alveolar macrophage redox state are reversible.

It has been demonstrated that hyperoxia can lead to changes in glutathione distribution and increased pentose shunt turnover in lungs [79]. In contrast to the reversible effects observed with the brief incubation of alveolar macrophages with paraquat or t-butyl hydroperoxide, prolonged

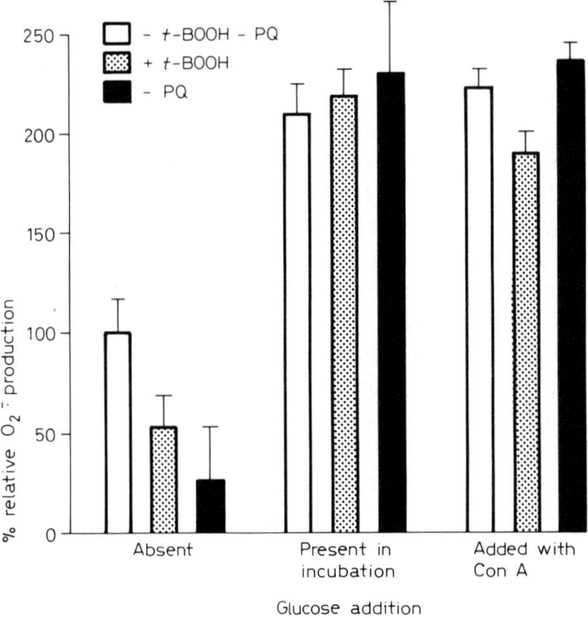

Fig. 6. Concanavalin A-stimulated superoxide production in the presence and absence of t-BOOH, paraquat (PQ), or glucose. Cells were incubated in the presence or absence of 5 mM glucose and in the presence or absence of either 10 μM t-BOOH or 1 mM paraquat. Following 15 min of incubation, the cells were then stimulated with 200 mg/ml concanavalin A in a 1-ml assay. Glucose was added with concanavalin A where indicated.

incubation of alveolar macrophage in hyperoxia, either in vivo or in vitro, causes irreversible inhibition of the respiratory burst [8, 10]. In our initial study of O_2-exposed alveolar macrophages [10], we measured the respiratory burst only in the presence of glucose. As a reversible decrease in the respiratory burst has also been noted in alveolar macrophages exposed in vitro to either paraquat or t-butyl hydroperoxide under conditions where sutstrate for the pentose phosphate pathway is limiting and NADPH depletion occurs, we subsequently examined whether there is also a glucose-dependent, reversible component in hyperoxide exposure [122]. We also further examined the irreversible component to determine whether it is due to irreversible alteration of status of NADPH and glutathione. Alveolar macrophages from 36-hour O_2-exposed rats showed a significant decrease in O_2^- production following stimulation with concanavalin A, whether glucose was absent or present. With O_2-exposed cells the respiratory burst in the presence of glucose was 5.1 times the rate minus glucose while with

Table 1. Effects of $10\,\mu M$ t-butyl hydroperoxide (t-BOOH) and glucose on glutathione distribution

Incubation		GSH + GSSG	GSSR	Extracellular glutathione	GSSG, % $(GSSG \times 100)$
glucose	t-BOOH				$(GSH + GSSG)$
+	−	1.02 ± 0.03	0.34 ± 0.06	0.11 ± 0.01	0.57 ± 0.19
−	−	1.06 ± 0.05	0.38 ± 0.05	0.10 ± 0.01	0.92 ± 0.11
−	+	0.78 ± 0.03^a	0.61 ± 0.05^a	0.18 ± 0.01^c	7.40 ± 1.58^d
+	+	1.09 ± 0.04	0.35 ± 0.04	0.12 ± 0.01	0.76 ± 0.19
(1 min)	+	1.14 ± 0.05	0.46 ± 0.04^b	0.17 ± 0.01^c	1.03 ± 0.27

[a] $p < 0.01$ versus all other incubations.
[b] $p < 0.01$ versus ($-$glucose $+$ t-BOOH) but not different significantly from other incubations.
[c] $p < 0.01$ versus incubations other than that also designated with superscript 'c'.
[d] $p < 0.001$ versus all other incubations.
Alveolar macrophages were incubated with or without t-BOOH for 15 min at 37 °C. Glucose was either absent or present ($5\,\text{m}M$) during incubation or added for just 1 additional minute of incubation. Results are expressed as GSH equivalents in $\text{nmol}/10^6$ cells (n = 3–12 determinations). From Sutherland et al. [121].

control cells the ratio was 2.1. This significant difference indicates that the inhibition of the respiratory burst by hyperoxic exposure can be partially reversed by glucose.

Nevertheless, the principal focus of this study was on that part of the inhibition of the burst that is irreversible. We picked 36 h as the time of exposure since this is approximately midway in the course of the path of irreversible inhibition (at 60 h inhibition is 100%) and is also early enough to avoid the influx of polymorphonuclear leukocytes into the lung. Cells were measured for total soluble glutathione, NADPH, NADP, ascorbate, and ATP levels after incubation with glucose (table 2). While glutathione, ATP, and total nucleotide levels were the same in both air and O_2-exposed cells, the redox state of the $NADPH/NADP^+$ couple was slightly less oxidized actually in the O_2-exposed alveolar macrophages. The decrease in ascorbate was due to dilution of the extracellular ascorbate concentration rather than oxidation. These results indicate that hyperoxia did not produce an irreversible change in the cell redox or energy state. While it is likely that the part of the hyperoxia-induced inhibition that could be reversed by glucose was also due to reversible NADPH depletion, the irreversible effect of hyperoxic stress was not due to an irreversible alteration of NADPH or glutathione metabolism in alveolar macrophages.

No drop in ATP concentration was observed in 36 h of in vivo exposure to hyperoxia of rat alveolar macrophages. In human and guinea

Table 2. Effect of 36 h of exposure to 100% O_2 on metabolites in alveolar macrophages

Exposure	NADPH + NADP nmol/10^6 cells	% as NADP	Ascorbate[a] nmol/10^6 cells	Glutathione[b] nmol/10^6 cells	ATP nmol/10^6 cell
Air	0.102 ± 0.006	25.4 ± 2.4	3.04 ± 0.23	1.44 ± 0.09	1.96 ± 0.06
O_2	0.093 ± 0.007	20.8 ± 1.3[c]	2.34 ± 0.14[d]	1.62 ± 0.08	2.02 ± 0.10
n	9	9	12	12	4

[a] Total ascorbate (reduced and oxidized).
[b] Total glutathione (reduced and oxidized).
[c] $p < 0.05$ versus air exposure.
[d] $p < 0.01$ versus air exposure.
Alveolar macrophages were obtained from rats exposed to air or 1 atmosphere absolute O_2 for 36 h. Values shown are the mean $\pm$ standard error for n measurements. From Sutherland et al. [122].

pig alveolar macrophages exposed in vitro to hyperoxia, there is a slow decline in ATP content, which became significant after 48 h of exposure [118]. This suggests that the change in ATP content may reflect later damage that occurs in the course of O_2 toxicity than is involved in the loss of the respiratory burst.

Superoxide production by alveolar macrophages can be stimulated by both the lectin, concanavalin A, and the calcium ionophore, A23187. With alveolar macrophages from air-controlled rats, maximal O_2^- production stimulated by concanavalin A is enhanced 2.0-fold by the presence of 1.3 mM extracellular Ca^{2+}. A23187 stimulation is completely dependent upon extracellular Ca^{2+} [123]. Alveolar macrophages were isolated from rats exposed to 1 atmosphere absolute O_2 for up to 48 h. With concanavalin A stimulation, there was a time-dependent decrease in respiratory burst activity in both the presence or absence of extracellular Ca^{2+}; however, the ratio of activity in the presence and absence of extracellular Ca^{2+} fell to 1.7 at 36 h and 1.1 at 48 h of O_2 exposure (fig. 7). With A23187 stimulation, O_2^- production declined 62% at 48 h. Using the fluorescent indicator, quin 2, we observed that changes in intracellular free Ca^{2+} following stimulation by concanavalin A were altered from the normal pattern in a similar way by either O_2 exposure or treatment with the Ca^{2+} entry blocker, verapamil (fig. 8). The results suggest that loss of the respiratory burst due to hyperoxia involves at least two separate components of the stimulus-response mechanism; one is a decreased accessibility of extracellular Ca^{2+} while the other is independent of Ca^{2+} entry from the external medium.

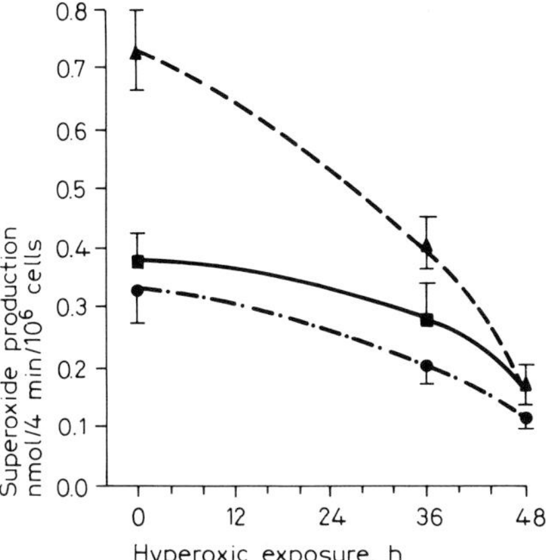

Fig. 7. Effect of hyperoxic exposure on the concanavalin A-stimulated respiratory burst of rat alveolar macrophages. Superoxide production was stimulated by addition of 250 mg/ml concanavalin A and measured as superoxide dismutase inhibitable cytochrome c reduction. Assay contained 0 extracellular calcium ($\blacksquare$) 1.3 mM extracellular calcium ($\blacktriangle$), or 1.3 mM extracellular calcium plus 20 μM verapamil ($\bullet$). From Forman et al. [123].

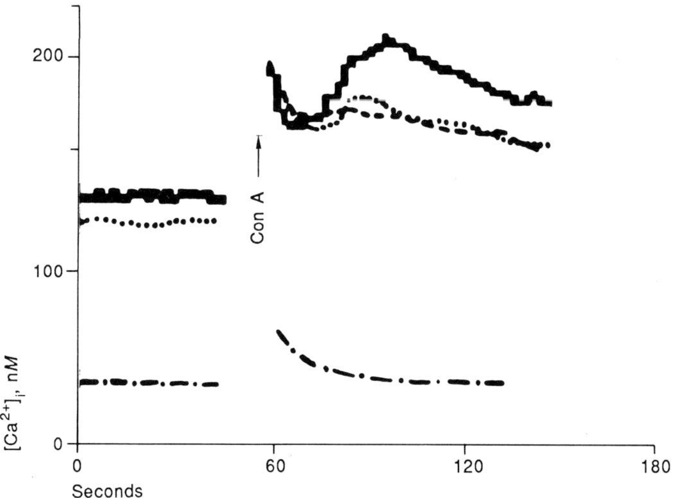

Fig. 8. Calculated average intracellular free calcium $[Ca^{2+}]_i$ in resting and concanavalin A-stimulated alveolar macrophages: Control (——), 48-hour oxygen exposed (– – –) verapamil (20 mM) treated ($\cdots$) and EGTA (2 mM) treated (–.–.–.) alveolar macrophages. Fluorescence measurements were made in the cells with quin 2. The cells were incubated for 15 min at 37 °C with 10 μM quin-2-methyl ester plux 1.3 mM Ca^{2+}. The cells were then washed and resuspended in buffer containing 1.3 mM Ca^{2+}. Verapamil, EGTA and O_2 exposure suppressed the second rise in $[Ca^{2+}]_i$; this is observed with control cells following concanavalin A stimulation. From Forman et al. [123].

Conclusion

The lung, because of its unique position as the interface with the atmosphere, is more susceptible to hyperoxia. Selective uptake of paraquat makes the pulmonary type II cell a target for damage by an agent, which can damage most cells at high concentrations.

Oxidant injury of cells is a consequence of an imbalance between the generation of O_2-derived free radicals and the antioxidant defenses, which nature has provided. Once the scales are tipped toward oxidation of cellular components, irreversible damage to cells will occur leading to loss of cellular function and, eventually, organ failure. A change in the redox state of the cell may permit damage to occur because of a decline in antioxidant capacity; however, such a change in the redox state is not an irreversible process. Damage to cells may be observed as loss of cell function before gross changes, such as non-viability. What remains to be solved is the sequence of events that occurs between the increase in free radical production and the loss of cell function, how loss of cell function leads to organ damage, such as alveolar edema, and what therapeutic methods can be devised to prevent this injury.

Acknowledgements

I would like to thank June Nelson, Eric Rotman, Gerald Harrison, Mark Posner, Dr. Thomas Aldrich, Dr. Mitchell Glass, Dr. Mark Sutherland and, especially, Dr. Aron Fisher for their contributions to the research cited in this chapter.

References

1 Bean JW: Effects of oxygen at increased pressure. Physiol Rev 1945;25:1.
2 Clark JM, Lambertsen CJ: Pulmonary oxygen toxicity: a review. Pharmacol Rev 1971;23:37–133.
3 Balentine JD: Experimental pathology of oxygen toxicity; in Jöbsis FF (ed): Oxygen and Physiological Function. Dallas, Professional Information Library, pp 311–378.
4 Gacad G, Massaro D: Hyperoxia: influence on lung mechanics and protein synthesis. J Clin Invest 1973;52:559.
5 Crapo JD, Barry BE, Foscue HA: Structural and biochemical changes in rat lungs occurring during exposures to lethal and adaptive doses of oxygen. Am Rev Respir Dis 1980;122:123–143.
6 Kistler GS, Caldwell PRB, Weibel ER: Development of fine structural damage to alveolar and capillary lining cells in oxygen poisoned rat lungs. J Cell Biol 1967;32:605–628.
7 Kapanci Y, Weibel ER, Kaplan HP, et al: Pathogenesis and reversibility of the pulmonary lesions of oxygen toxicity in monkeys. II. Ultrastructural and morphometric studies. Lab. Invest. 1969;20:101.

 8 Rister M, Baehner RL: Effect of hyperoxia on superoxide anion and hydrogen peroxide
    production of polymorphonuclear leukocytes and alveolar macrophages. Br J Haematol.
    1977;36:241–248.
 9 Wolff LJ, Boxer LA; Allen JM: et al: The selective effect of hyperoxia on the guinea pig
    alveolar macrophage membrane. J Reticuloendothel Soc 1978;24:377–382.
10 Forman HJ, Williams JJ, Nelson J, et al: Hyperoxia inhibits stimulated superoxide
    release by rat alveolar macrophages. J Appl Physiol 1982;53:685–689.
11 Suttorp N, Simon LM: Decreased bactericidal function and impaired respiratory burst
    in lung macrophages after sustained in vitro hyperoxia. Am Rev Respir Dis
    1983;128:486–490.
12 Adamson IVR, Bowden DH: The type II cells as progenitor of alveolar epithelial
    regeneration. Lab Invest 1974;30:35–42.
13 Crapo JD, Peters-Golden M, Marsh-Salin J, et al: Pathologic changes in the lungs of
    oxygen-adapted rats. A morphometric analysis. Lab Invest 1978;39:640–653.
14 Fox RB, Hoidal JR, Brown DM, et al: Pulmonary inflammation due to oxygen toxicity:
    involvement of chemotactic factors and polymorphonuclear leukocytes. Am Rev Respir
    Dis 1981;123:521–523.
15 Baehner RL, Boxer LA, Higgins C, et al: Accumulation of circulating macrophages in
    lungs of guinea pigs exposed to hyperoxia. Pediatr. Res.1981;15:1356–1358.
16 Fisher AB, Forman HJ, Glass M: Mechanisms of pulmonary oxygen toxicity. Lung
    1984;162:255–259.
17 Oreopoulos DE, Soyannwo MAO, Sinniah R, et al: Acute renal failure in case of
    paraquat poisoning. Br Med J 1968;749–750.
18 Waddell WJ, Marlowe C: Tissue and cellular distribution of paraquat in mice. Toxicol
    Appl Pharmacol 1980;56:127–140.
19 Smith P, Heath D, Kay JM: The pathogenesis and structure of paraquat-induced
    pulmonary fibrosis in rats. J Pathol 1974;114:57–67.
20 Deneke SM, Fanburg BL: Oxygen toxicity of the lung: an update. Br J Anaesth
    1982;54:737–749.
21 Deneke SM, Fanburg BL: Normobaric oxygen toxicity of the lung. N Engl J Med
    1980;303:76–86.
22 Forman HJ, Fisher AB: Anti-oxidant defenses; in Gilbert DL (ed): Oxygen and Living
    Processes. Berlin, Springer, 1981, pp 237–249.
23 Freeman BA, Crapo JD: Free radicals and tissue injury. Lab Invest 1982;47:412–426.
24 Frank L, Massaro D: Oxygen toxicity. Am J Med 1980;69:117–126.
25 Mustafa MG, Tierney DF: Biochemical and metabolic changes in the lung with oxygen,
    ozone, and nitrogen dioxide toxicity. Am Rev Respir Dis 1978;118:1061–1090.
26 Fridovich I: The biology of oxygen radicals. Science 1978;201:875–880.
27 Fridovich I: Superoxide radical and superoxide dismutases; in Gilbert DL (ed): Oxygen
    and Living Processes. Berlin, Springer, 1981, pp 250–272.
28 Taube H: Mechanisms of oxidation with oxygen. J Gen Physiol 1965;2:29.
29 Boveris A, Oshino M, Chance B: The cellular production of hydrogen peroxide.
    Biochem J 1972;128:617.
30 Oshino N, Chance B, Sies H, et al: The role of H_2O_2 generation in perfused rat liver and
    the reaction of catalase compound I and hydrogen donors. Arch Biochem Biophys
    1973;154:117.
31 Fridovich I, Handler P: Xanthine oxidase. V. Differential inhibition of the reduction of
    various electron acceptors. J Biol Chem 1962;237:916–921.
32 Fridovich I: in Hayaishi O (ed): Molecular Mechanisms of Oxygen Activation. New
    York, Academic Press, 1974, pp 453–477.

33 Boveris A, Chance B: Optimal rates of hydrogen peroxide production in hyperbaric oxygen; in Thurman RG, et al (eds): Alcohol and Aldehyde Metabolizing Systems. New York, Academic Press, 1977, pp 207–214.

34 Freeman BA, Crapo JD: Hyperoxia increases oxygen radical production in rat lungs and lung mitochondria. J Biol Chem 1981;256:10986–10992.

35 Turrens JF, Freeman BA, Crapo JD: Hyperoxia increases H_2O_2 release by lung mitochondria and microsomes. Arch Biochem Biophys 1982;217:411–421.

36 Forman HJ, Boveris A: Superoxide radical and hydrogen peroxide in mitochondria; in Pryor WA (ed): Free Radicals in Biology. New York, Academic Press, 1982, vol V, pp 65–90.

37 Yusa T, Crapo JD, Freeman BA: Hyperoxia enhances lung and liver nuclear superoxide generation. Biochim Biophys Acta 1984;798:167–174.

38 Rose MS, Smith LL: Tissue uptake of paraquat and diquat. Gen Pharmacol 1977;8:173–176.

39 Forman HJ, Aldrich TK, Posner MA, et al: Differential paraquat uptake and redox kinetics of rat granular pneumocytes and alveolar macrophages. J Pharmacol Exp Ther 1982;221:428–433.

40 Farrington JA, Ebert M, Land EJ, et al: Bipyridylium quaternary salts and related compounds. V. Pulse radiolysis studies of the reaction of paraquat radical with oxygen. Implication for the mode of action of bipyridyl herbicides. Biochem Biophys Acta 1973;314:372–381.

41 Harper TW, Westcott JY, Voelkel N, et al: Metabolism of leukotrienes B_4 and C_4 in the isolated perfused rat lung. J Biol Chem 1984;259:14437–14440.

42 Pedersen TC, Aust SD: The role of superoxide and singlet oxygen in lipid peroxidation promoted by xanthine oxidase. Biochem Biophys Res Commun 1973;52:1071–1078.

43 Tien M, Svingen BA, Aust SD: Superoxide dependent lipid peroxidation. Fed Proc 1981;40:179–182.

44 Hochstein P, Ernster L: ADP-activated lipid peroxidation coupled to the TPNH oxidase system of microsomes. Biochem Biophys Res Commun 1963;12:388–394.

45 Poyer JL, McCay PB: Reduced triphosphopyridine nucleotide oxidase-catalyzed alteration of membrane phospholipids. IV. Dependence on Fe^{3+}. J Biol Chem 1971;246:263–269.

46 Bielski BHJ, Arudi RL, Sutherland MW: A study of the reactivity of HO_2/O_2^- with unsaturated fatty acids. J Biol Chem 1983;258:4759–4761.

47 Taylor L, Menconi MJ, Polgar P: The participation of hydroperoxides and oxygen radicals in the control of prostaglandin synthesis. J Biol Chem 1983;258:6855–6857.

48 Marshall PJ, Kulmacz RJ, Lands WEM: Hydroperoxides, free radicals and prostaglandin synthesis, in Bors W, Saran M, Tait D (eds): Oxygen Radicals in Chemistry and Biology. Berlin, de Gruyter, 1984.

49 Haugaard N: Cellular mechanisms of oxygen toxicity. Physiol Rev 1968;48:311.

50 Levine RL: Oxidative modification of glutamine synthetase. I. Inactivation is due to loss of one histidine residue. J Biol Chem 1983;258:11823–11827.

51 Forman HJ, Evans HJ, Hill RL, et al: Histidine at the active site of superoxide dismutase. Biochemistry 1973;12:823–827.

52 Hodgson EK, Fridovich I: The interaction of bovine erythrocyte superoxide dismutase with hydrogen peroxide: inactivation of the enzyme. Biochemistry 1975;14:5294–5299.

53 Bray RC, Cockle SH, Fielden EM, et al: Reduction and inactivation of superoxide dismutase by hydrogen peroxide. Biochem J 1974;139:43.

54 Kono Y, Fridovich I: Superoxide radical inhibits catalase. J Biol Chem 1982;257:5751–5754.

55 Rister M, Baehner RL: The alteration of superoxide dismutase, catalase, glutathione peroxidase, and NAD(P)H cytochrome c reductase in guinea pig polymorphonuclear leukocytes and alveolar macrophages during hyperoxia. J Clin Invest 1976;58:1174–1184.

56 Chance B, Jamieson D, Coles H: Energy-linked pyridine nucleotide reduction: inhibitory effects of hyperbaric oxygen in vitro and in vivo. Nature 1965;206:257.

57 Allen JE, Goodman DBP, Besarab A, et al: Studies on the biochemical basis of oxygen toxicity. Biochim Biophys Acta 1973;320:708–728.

58 Falsetti H: Effect of oxygen tension on sodium transport across isolated frog skin. Proc Soc Exp Biol Med 1959;101:721–725.

59 Sanders AP, Currie WD: Chemical protection against oxygen toxicity; in Lambersten CJ (ed): Underwater Physiology. 5th Symp on Underwater Physiology, pp 35–40.

60 Aerts C, Voisin C: In vitro toxicity of oxygen and oxygen-paraquat association on alveolar macrophages surviving in gas phase. Bull Eur Physiopathol Respir 1981;17(suppl):145–151.

61 Block ER, Fisher AB: Depression of serotonin clearance by rat lungs during oxygen exposure. J Appl Physiol 1977;42:33–38.

62 Dobuler KJ, Catravas JD, Gillis CN: Early detection of oxygen-induced lung injury in conscious rabbits. Reduced in vivo activity of angiotensin converting enzyme and removal of 5-hydroxytryptamine. Am Rev Respir Dis 1982;126:534–539.

63 Lotscher HR, Winterhalter KH, Carafoli E, et al: Hydroperoxides can modulate the redox state of pyridine nucleotides and the calcium balance in rat liver mitochondria. Proc Natl Acad Sci USA 1979;76:4340–4344.

64 Lotscher HR, Winterhalter KH, Carafoli E, et al: Hydroperoxide-induced loss of pyridine nucleotides and release of calcium from rat liver mitochondria. J Biol Chem 1980;255:9325–9330.

65 Jones DP, Thor H, Smith MT, et al: Inhibition of ATP-dependent microsomal Ca^{2+} sequestration during oxidative stress and its prevention by glutathione. J Biol Chem 1983;258:6390–6393.

66 Bellomo G, Jewell SA, Thor H, et al: Regulation of intracellular calcium compartmentation: Studies with isolated hepatocytes and t-butyl hydroperoxide. Proc Natl Acad Sci USA 1982;79:6842–6846.

67 Lehninger AL, Vercesi A, Bababunmi EA: Regulation of Ca^{2+} release from mitochondria by the oxidation-reduction state of pyridine nucleotides. Proc Natl Acad Sci USA 1978;75:1690–1694.

68 Beatrice MC, Palmer JW, Pfeiffer DR: The relationship between mitochondrial membrane permeability, membrane potential, and the retention of Ca^{2+} by mitochondria. J Biol Chem 1980;255:8663.

69 Scarpa A: in Grebisch G, et al (eds): Membrane Transport in Biology. New York, Springer, 1979, p 263.

70 Black BL, Jarrett L, MacDonald JM: The regulation of endoplasmic reticulum calcium uptake of adipocytes by cytoplasmic calcium. J Biol Chem 1981;256:322.

71 Jakoby WB, Keen JH: A triple threat in detoxification: the glutathione-S-transferases. Trends Biochem Sci 1977;2:229.

72 Tierney D, Ayres L, Herzog S, et al: Pentose pathway and production of reduced nicotinamide adenine dinucleotide phosphates: a mechanism that may protect the lung from oxidants. Am Rev Respir Dis 1973;108:1348–1351.

73 Crapo JD, Tierney DF: Superoxide dismutase and pulmonary oxygen toxicity. Am J Physiol 1974;226:1401–1407.

74 Crapo JD, Sjostrom K, Drew RT: Tolerence and cross-tolerance using NO_2 and O_2 I. Toxicology and biochemistry. J Appl Physiol 1978;44:364–369.

75 Forman HJ, Fisher AB: Antioxidant enzymes in rat granular pneumocytes: constitutive levels and effect of hyperoxia. Lab Invest 1981;45:1–6.

76 Stevens JB, Autor AP: Oxygen-induced synthesis of superoxide dismutase and catalase in pulmonary macrophages of neonatal rats. Lab Invest 1977;37:470–478.

77 Autor AP, Stevens JB: Mechanism of oxygen detoxification in neonatal rat lung tissue. Photochem Photobiol 1978;28:775–780.

78 Stevens JB, Autor AP: Induction of superoxide dismutase by oxygen in neonatal rat lung. J Biol Chem 1977;252:3509–3514.

79 Kimball RE, Reddy K, Peirce TH, et al: Oxygen toxicity: augmentation of antioxidant defense mechanisms in rat lung. Am J Physiol 1976;230:1425–1431.

80 Crapo JD, McCord JM: Oxygen-induced changes in pulmonary superoxide dismutase assayed by antibody titration. Am J Physiol 1976;231:1196–1203.

81 Frank L, Bucher JR, Roberts RJ: Oxygen toxicity in neonatal and adult animals of various species. J Appl Physiol 1978;45:699–704.

82 Frank L, Yam J, Roberts RJ: The role of endotoxin in protection of adult rats from oxygen-induced lung toxicity. J Clin Invest 1978;61:269–275.

83 Hass MA, Frank L, Massaro D: The effect of bacterial endotoxin on synthesis of (Cu, Zn) superoxide dismutase in lungs of oxygen-exposed rats. J Biol Chem 1982;257:9379–9383.

84 Frank L, Summerville J, Massaro D: Protection from oxygen toxicity with endotoxin. Role of the endogenous antioxidant enzymes of the lung. J Clin Invest 1980;65:1104–1110.

85 Ody C, Bach-Dieterle Y, Wand I, et al: Effect of hyperoxia on superoxide dismutase content of pig pulmonary artery and aortic endothelial cells in culture. Exp Lung Res 1980;1:271–279.

86 Jenkinson SG, Lawrence RA, Burk RF, et al: Non-selenium-dependent glutathione peroxidase activity in rat lung: association with lung glutathione S-transferase activity and the effects of hyperoxia. Toxicol Appl Pharmacol 1983;68:399–404.

87 Yam J, Frank L, Roberts RJ: Oxygen toxicity; comparison of lung biochemical responses in neonatal and adult rats. Pediatr Res 1978;12:115–119.

88 Freeman BA, Young SL, Crapo JD: Liposome-mediated augmentation of superoxide dismutase in endothelial cells prevents oxygen injury. J Biol Chem 1983;258:12534–12542.

89 Turrens JF, Crapo JD, Freeman BA: Protection against oxygen toxicity in intravenous injection of liposome-entrapped catalase and superoxide dismutase. J Clin Invest 1984;73:87–95.

90 Simon LM, Liu J, Theodore J, et al: Effect of hyperoxia, hypoxia, and maturation on superoxide dismutase activity in isolated alveolar macrophages. Am Rev Respir Dis 1977;115:279–284.

91 Deneke SM, Bernstein SP, Fanburg BL: Enhancement by disulfiram (antabuse) of toxic effects of 95 to 97% O_2 on the rat lung. J Pharmacol Exp Ther 1978;208:377–380.

92 Forman HJ, York JL, Fisher AB: Mechanism for the potentiation of oxygen toxicity by disulfiram. J Pharmacol Exp Ther 1980;212:452–455.

93 Forman HJ, Rotman EI, Fisher AB: The roles of selenium and sulfur-containing amino acids in protection against oxygen toxicity. Lab Invest 1983;49:148–153.

94 Cross CE, Hasegawa G, Reddy KA, et al: Enhanced lung toxicity of O_2 in selenium-deficient rats. Res Commun Chem Pathol Pharmacol 1977;16:695–706.

95 Deneke SM, Gershoff SN, Fanburg BL: Potentiation of oxygen toxicity in rats by dietary protein or amino acid deficiency. J Appl Physiol 1983;54:147–151.

96   Rosenbaum RM, Wittner M, Lenger M: Mitochondrial and other ultrastructural changes in great alveolar cells of oxygen-adapted and poisoned rats. Lab Invest 1969;20:516–528.

97   Sevanian A, Davies KJA, Hochstein P: Conservation of vitamin C by uric acid in blood. J Free Radicals Biol Med 1985;1:117–124.

98   Aldrich TK, Fisher AB, Forman HJ: Paraquat inhibits mixed-function oxidation by rat lung. J Appl Physiol 1983;54:1089–1093.

99   Aldrich TK, Fisher AB, Cadenas E, et al: Evidence for lipid peroxidation by paraquat in the perfused rat lung. J Lab Clin Med 1983;101:66–73.

100   Babior BM: Oxygen-dependent microbial killing by phagocytes. N Engl J Med 1978;298:659–668.

101   Kazmierowski JA, Gallin JI, Reynolds J: Mechanism of the inflammatory response in primate lungs. Demonstration and partial characterization of an alveolar macrophage-derived chemotactic factor with preferential activity for polymorphonuclear leukocytes. J Clin Invest 1977;59:273–281.

102   Gadek JE, Hunninghake GW, Zimmerman RL, et al: Regulation of the release of alveolar macrophage-derived neutrophil chemotactic factor. Am Rev Respir Dis 1980;121:723–733.

103   Hseuh W, Sun FF: Leukotriene B_4 biosynthesis by alveolar macrophages. Biochem Biophys Res Commun 1982;106:1085–1091.

104   Johnston RB Jr, Godzik CA, Cohn ZA: Increased superoxide anion production by immunologically activated and chemically elicited macrophages. J Exp Med 1978;148:115–127.

105   Cameron AR, Nelson J, Forman HJ: Depolarization and increased conductance precede superoxide release by concanavalin A stimulated rat alveolar macrophages. Proc Natl Acad Sci USA 1983;80:3726–3728.

106   Lew PD, Stossel TP: Effect of calcium on superoxide production by phagocytic vesicles from rabbit alveolar macrophages. J Clin Invest 1981;67:1–9.

107   Forman HJ, Nelson J: Effect of extracellular calcium on superoxide release by rat alveolar macrophages. J Appl Physiol 1983;54:1249–1253.

108   Joseph SK: Inositol trisphosphate: an intracellular messenger produced by Ca^{2+} mobilizing hormones. Trends Biochem Sci 1984;9:420–421.

109   Berridge MJ, Irvine RF: Inositol trisphosphate, a novel second messenger in cellular signal transduction. Nature 1984;312:315–321.

110   Hirata M, Suematsu E, Hashimoto T, et al: Release of Ca^{2+} from a non-mitochondrial store site in peritoneal macrophages treated with saponin by inositol 1,4,5-triphosphate. Biochem J 1984;223:229–236.

111   Stickle DF, Daniele RP, Holian A: Cytosolic calcium, calcium fluxes, and regulation of alveolar macrophage superoxide anion production. J Cell Physiol 1984;121:458–466.

112   Cohen HJ, Chovaniec ME: Superoxide generation by digitonin-stimulated guinea pig granulocytes. A basis for a continuous assay for monitoring superoxide production and for the study of the activation of the generating system. J Clin Invest 1981;61:1081–1087.

113   Forman HJ, Nelson J, Fisher AB: Rat alveolar macrophages require NADPH for superoxide production in the respiratory burst. J Biol Chem 1980;255:9879–9883.

114   Simon LM, Axline SG, Robin ED: The effect of hyperoxia on phagocytosis and pinocytosis in isolated pulmonary macrophages. Lab Invest 1978;39:541–546.

115   Raffin TA, Simon LM, Braun D, et al: Impairment of phagocytosis by moderate hyperoxia (40%–60% oxygen) in lung macrophages. Lab Invest 1980;42:622–626.

116   Johnson GS, Rister M, Higgins C, et al: Biochemical basis of oxygen toxicity in guinea pig alveolar macrophages and granulocytes. 16th Annu Hanford Biol. Symp, pp 509–522.

117 Rister M: Effects of hyperoxia on phagocytosis. Blut 1982;45:157–166.
118 Voisin C, Aerts CF, Tonnel AB: Application à l'étude de la cytotoxicité de l'oxygène. Colloq INSERM 1979;84:169–176.
119 Davis WB, Rennard SI, Bitterman PB, et al: Pulmonary oxygen toxicity: early reversible changes in human alveolar structures induced by hyperoxia. Engl J Med 1983;309:878–883.
120 Bus JS, Aust SD, Gibson JE: Superoxide- and singlet oxygen-catalyzed lipid peroxidation as a possible mechanism for paraquat (methyl viologen) toxicity. Biochem Biophys Res Commun 1974;58:749.
121 Sutherland MW, Nelson J, Harrison G, et al: Effects of t-butyl hydroperoxide on NADPH, glutathione, and respiratory burst of rat alveolar macrophages. Arch Biochem Biophys 1985;243:325–331.
122 Sutherland MW, Glass M, Nelson J, et al: Oxygen toxicity: loss of lung macrophage function without metabolite depletion. J Free Radicals Biol Med 1985;1:209–214.
123 Forman HJ, Nelson J, Harrison G: Hyperoxia alters the effect of Ca^{2+} on rat alveolar macrophage superoxide production. J Appl Physiol 1986;60:1300–1305.

Henry Jay Forman, PhD, University of Southern California,
Children's Hospital of Los Angeles, Los Angeles, CA 90027 (USA)

4 Mechanism of Free Radical Generation during Reperfusion of Ischemic Myocardium

Dipak K. Das, Richard M. Engelman

Myocardial cellular injury associated with the reperfusion of ischemic myocardium has been attributed to many interrelated factors. One of these factors, which is not yet highly explored, is free radical mechanisms. The involvement of superoxide radicals in reperfusion injury is indeed an interesting facet of the pathobiology of myocardial injury. The participation of the free radicals has been hypothesized from the beneficial effects of anitoxidative enzymes [1] and free radical scavengers [2–10] on myocardial function. In addition, the consequence of free radical attack has been demonstrated by identifying the lipid peroxidative products in reoxygenated myocardium [11–14]. More recently, free radicals are directly identified in the ischemic and reperfused heart [15–17].

The above studies, although limited in number, strongly support the free radical mechanism in ischemic reperfusion injury. There is no question that free radicals are involved, but their origin is unknown. What triggers the generation of these radicals? Where do the signals originate and where do the attacks begin? These and many more questions remain unanswered. The intent of this review is to provide an overview of the possible mechanism that might be involved in the generation of free radicals during reperfusion of ischemic myocardium. Several mechanisms can be hypothesized based upon currently available knowledge in this particular research area. There can be several factors, one of which may act as second messenger in triggering the radical attack. The reactive oxygen-derived free radicals are quite capable of oxidizing biomolecules such as the membrane proteins or phospholipids leading to ultimate tissue damage. Finally, based on our recent study in conjunction with other studies, a novel mechanism is proposed and some evidence presented to warrant the validity of this hypothesis.

Sources of Free Radicals

Activation of Polymorphonuclear Leukocytes

Reperfusion of ischemic myocardium was performed with blood containing polymorphonuclear leukocytes (PMN). Studies have been carried out which show that these PMN cross the endothelial membrane of the vascular bed into the ischemic myocardium [7, 18], possibly mediated by chemotactic factors [9–21]. These PMN may be activated and then possess the ability to produce superoxide radicals [22–29]. Virtually all of the oxygen consumed by PMN during the respiratory burst is reduced to superoxide via a single electron transfer reaction. The superoxide anion (O_2^-) generated by these stimulated phagocytes undergoes a series of secondary reactions, producing H_2O_2 [25] and hydroxyl radicals (OH') and HOCL [27]. The production of the superoxide anion is catalyzed by a membrane-bound, NADPH-dependent flavoprotein oxidase [29] which is abundantly present in the myocardium [29]. It has been shown that only activated PMN are capable of producing the superoxide radicals [30], and this activation can occur with the presence of complement fragments in the system [31–33]. Such complement gains access to the myocardial extravascular space by crossing the endothelial membrane of the vascular tree [19]. A second potential activating agent for PMN is arachidonic acid [34], which accumulates in the ischemic myocardium associated with inhibition of β-oxidation [35] and defective reacylation [36, 37] during ischemia. In addition, several catabolic products of arachidonic acid generated via the lipooxygenase pathway are also known the be chemotactic [21].

Studies supporting the role of the PMN in reperfusion injury are mostly indirect rather than direct measurements, e.g. a non-steroidal anti-inflammatory agent, ibuprofen, has been found to inhibit the PMN influx into the infarcted tissue with a corresponding reduction of ultimate infarct size [19]. Similarly, PMN depletion was associated with a significant reduction in ultimate infarct size in experimental animals [18].

Other studies, however, showed no decrease in infarct size in leukopenic dogs after 21 h of reperfusion [38]. A recent sutdy from the authors' laboratories also indicated that despite the inhibition of PMN and platelet deposition in the ibuprofen-treated pig heart, myocardial reperfusion injury was increased when compared to controls. These studies demonstrated decreased adenosine triphosphate (ATP) and creatine phosphate (CP) levels with a corresponding increase in creatine kinase (CK) release in the ibuprofen-treated animals after 60 min of reperfusion (table 1). The poor recovery of ibuprofen-treated hearts may be explained by one or more reasons. Ischemic and reperfused hearts are known to accumulate significant amounts of free fatty acids, especially arachidonic acid, as a result of the breakdown of membrane phospholipids [36, 37]. The accumulated arachidonic acid in ischemic heart cannot be reincorporated into the membrane phospholipids due to defective reacylation [37], but can serve as substrate for the cyclooxygenase pathway, thereby enhancing prostacyclin (PGI_2) and thromboxane production [39]. These studies support the previous observations and further demonstrate that inhibition of the cyclo-

Table 1. Effects of ibuprofen on high energy phosphate compounds and arachidonic acid metabolites in the heart during ischemia and reperfusion

		Post-IBU pre-ischemia	60-min occlusion	120-min occlusion	60-min reperfusion
ATP	Cont	4.43 ± 0.10	0.81 ± 0.09	0.54 ± 0.06	1.34 ± 0.32
μmol/g	IBU	4.75 ± 0.54	1.66 ± 1.19	0.38 ± 0.9	$0.35 \pm 0.10*$
CP	Cont	7.22 ± 0.46	1.86 ± 0.43	0.95 ± 0.21	5.33 ± 0.53
μmol/g	IBU	9.83 ± 0.90	3.63 ± 2.60	0.77 ± 0.26	$1.26 \pm 0.29*$
CK release	Cont	92.8 ± 12.3	96.1 ± 10.9	–	238.5 ± 9.4
% control	IBU	83.5 ± 6.2	93.7 ± 8.8	–	377.2 ± 35.8
Arachidonic	Cont	18 ± 3	42 ± 5	61 ± 7	98 ± 5
acid $C_{20:4}$,	IBU	11 ± 5	$68 \pm 11*$	$103 \pm 14*$	$230 \pm 23*$
nmol/g					

ATP = Adenosine triphosphate; CP = creatine phosphate; CK = creatine kinase.
* $p < 0.05$ comparing animals with and without ibuprofen.

oxygenase pathway by ibuprofen results in additional accumulation of arachidonic acid. The polyunsaturated fatty acids can harm the heart in several ways. Free fatty acids, by their detergent properties, can disrupt the cell membrane. In addition, they may inhibit some important enzyme systems for the Na/K pump [35]. Arachidonic acid is also chemotactic and can act as an ionophoretic agent causing a massive influx of Ca^{2+} ions.

Furthermore, free radicals are being generated in the ischemic-reperfused heart which attack polyunsaturated fatty acids, such as arachidonic acid, causing lipid peroxidation [12, 13].

This study also demonstrated a statistically significant decrease in arachidonate metabolites (6-keto-prostaglandin $F_{1\alpha}$ and thromboxane B_2) in the ibuprofen-treated group when compared to controls (fig. 1, 2). This could also explain the poor recovery of the ibuprofen-treated animals after 60 min of reperfusion. Exogenously administered vasodilatory PGI_2 has recently been shown to increase blood flow to ischemic myocardium, improve myocardial segment shortening, and decrease infarcy size. In addition, PGI_2 posesses anti-aggregatory properties, and decrease in PGI_2 formation presumably favors platelet adhesion to the already injured myocardium because it has been shown that while small concentrations of PGI_2 prevent thrombus formation, larger concentrations are required to reduce platelet adhesion [40]. By inhibiting vascular PGI_2 production through ibuprofen administration, continued myocardial ischemia may occur during the reperfusion period.

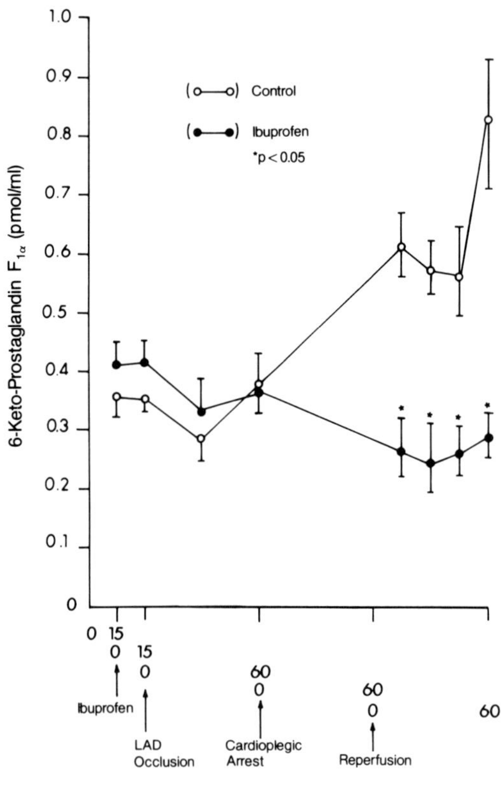

Fig. 1. Time course of change in the level of 6-keto-prostaglandin $F_{1\alpha}$ (6-keto-PGF$_{1\alpha}$) in the heart as a function of time: duration of left anterior descending coronary artery (LAD) occlusion, arrest, and reperfusion. Isolated in situ pig hearts were preincubated for 15 min with/without ibuprofen and then subjected to 60 min of LAD occlusion, followed by 60 min of hypothermic cardioplegic arrest and 60 min of reperfusion. ○ = Control group; ● = ibuprofen-treated group. Results are expressed as means ± SE of 6 separate experiments for control and 6 experiments for the ibuprofen-treated group.

Finally, by blocking the cyclooxygenase pathway, ibuprofen can potentiate generation of leukotrienes via the lipooxygenase pathway [41]. These leukotrienes are well-known mediators for vasoconstriction and can secondarily decrease myocardial contractility. Previous studies demonstrated that intracoronary administration of leukotrienes caused a decrease in both coronary blood flow and global myocardial contractility [42]. Leukotrienes also are known to possess phlogistic activity and can serve as

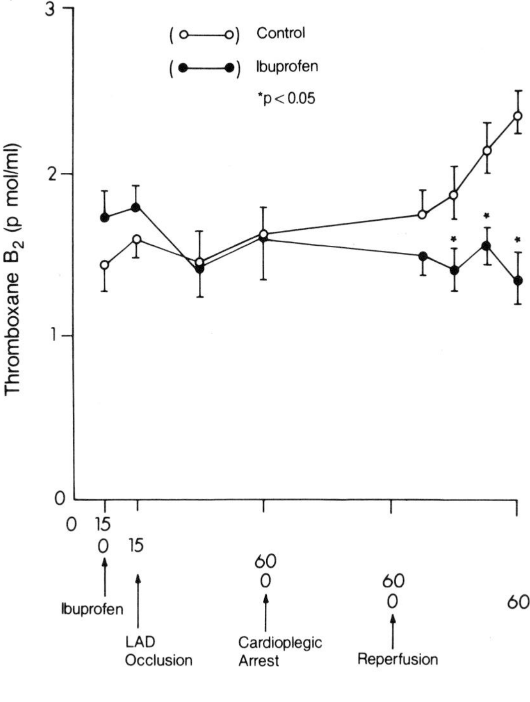

Fig. 2. Time courses of change in the level of thromboxane B$_2$ (ThB$_2$) in the heart as a function of time: duration of left anterior descending coronary artery (LAD) occlusion, arrest, and reperfusion. Isolated in situ pig hearts were preincubated for 15 min with/without ibuprofen and then subjected to 60 min of LAD occlusion, followed by 60 min of hypothermic cardioplegic arrest and 60 min of reperfusion. $\bigcirc$ = Control group; $\bullet$ = ibuprofen-treated group. Results are expressed as means $\pm$ SE of 6 separate experiments for control and 6 experiments for the ibuprofen treated group.

mediators of both the humoral and cellular phase of the inflammatory response [43].

In conclusion, inhibition of PMN and platelet influx into the ischemic-reperfused myocardium by a nonsteroidal anti-inflammatory agent such as ibuprofen failed to mitigate reperfusion injury in the isolated in situ pig heart, primarily because of the inhibitory effect of ibuprofen on the cyclooxygenase pathway. Arachidonic acid accumulation appears to be more detrimental than its metabolism via the cyclooxygenase pathway. The results also demonstrate that the presence of PGI$_2$ in ishcemic heart is far more beneficial than its absence.

Table 2. Effect of trifluoperazine on coronary blood flow and myocardial contractility and compliance during ischemia and reperfusion

		Post-TFP pre-ischemia	60-min LAD occlusion	15-min reperfusion	60-min reperfusion
Coronary	Cont	96.1 ± 5.6	69.5 ± 6.9	53.0 ± 6.6	37.0 ± 4.9
blood flow	TFP	12.49 ± 9.9	96.7 ± 7.1	$104.3 \pm 20.4**$	$97.3 \pm 9.1**$
$LV\frac{dp}{dt}$	Cont	101.1 ± 7.8	56.5 ± 6.6	44.6 ± 3.7	28.6 ± 3.7
	TFP	83.7 ± 7.1	72.2 ± 13.4	56.2 ± 12.4	$56.1 \pm 11.4*$
LVDP	Cont	95.1 ± 2.2	61.2 ± 3.9	44.6 ± 2.9	27.9 ± 3.8
	TFP	89.3 ± 7.6	69.7 ± 11.3	71.7 ± 7.6	$73.0 \pm 13.9*$
LVEDP	Cont	12.8 ± 3.2	33.5 ± 3.6	45.9 ± 3.1	72.1 ± 3.9
	TFP	13.3 ± 3.4	13.9 ± 1.8	$24.3 \pm 3.3**$	$29.7 \pm 3.1**$
ΔL	Cont	100.6 ± 1.8	2.2 ± 1.4	20.4 ± 4.2	11.8 ± 2.6
	TPF	95.4 ± 5.3	4.0 ± 4.0	42.8 ± 6.9	$42.0 \pm 8.1*$

LV dp/dt and LVDP = LV developed pressure (all % control); LVEDP = left ventricular end-diastolic pressure (mm Hg) and ΔL = LV segmental fiber length shortening in ischemic region (% control).
* $p < 0.05$; ** $p < 0.01$.

Xanthine Oxidase-Xanthine Dehydrogenase System

Xanthine oxidase (XO) historically was the first documented biologic source of superoxide anions [44]. Various organs contain this enzyme [20] although current knowledge suggests that the in vivo form of the enzyme is not the superoxide producing oxidase (type O):

$$\text{Xanthine} + H_2O + 2O_2 \xrightarrow{\text{type O}} \text{Uric acid} + 2O_2 + 2H^+$$

but rather an NAD^+-reducing dehydrogenase (type D):

$$\text{Xanthine} + H_2O + NAD \xrightarrow{\text{type D}} \text{Uric acid} + NADH + H^+$$

It has been suggested that during ischemia, the xanthine dehydrogenase is rapidly converted to the oxidase form as a result of sulf-hydryl oxidation of limited proteolysis [45]. This type O form is not reversible by thiol reductants to type D, suggesting a proteolytic conversion. The rapidity of the conversion, however, cannot be explained. It seems too fast to be the result of lysosomal breakdown or to be linked to the development of severe acidosis within the cell. Only recently, however, has it been noted that calcium-regulated proteases are present within myocardial cells [46]. Because energy change drops very quickly in the ischemic myocardium, and because

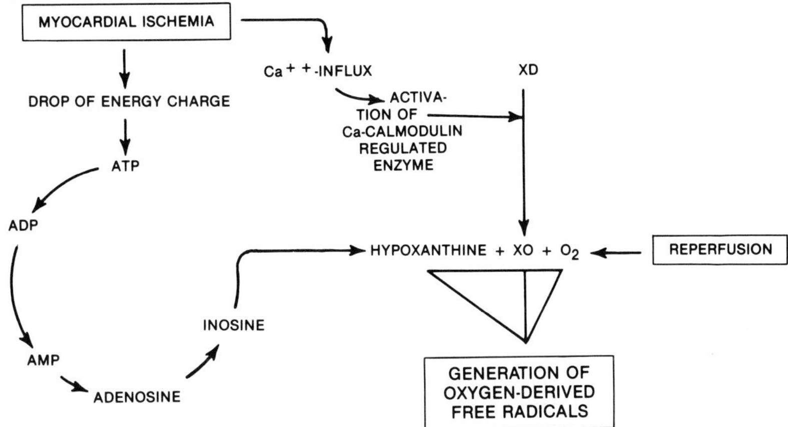

Fig. 3. Proposed model for the free-radical generation during reoxygenation by xanthine oxidase-hypoxanthine reaction.

maintenance of cation gradients is energy-dependent, it is conceivable that ischemia results in an influx of Ca^{2+} which, in turn, may activate the calcium-calmodulin-regulated enzymes. Indeed, the use of calcium-calmodulin antagonist has effectively reduced reperfusion injury [47]. It has been shown that Ca^{2+} channel blockers do not inhibit Ca^{2+} accumulation, nor do they protect ishcemic hearts from reperfusion injury when used prior to or during reperfusion. The use of calmodulin antagonists such as W-7 at the onset of reperfusion is equally effective as they are used before the ischemic insult to prevent Ca^{2+} accumulation as well as reperfusion injury [48].

Trifluoperazine, a clinically used antipsychotic drug, salvaged myocardial dysfunciton during ischemia and reperfusion by virtue of its ability to antagonize the calcium-calmodulin complex [49]. Trifluoperazine enhanced systolic segment shortening and improved myocardial contractility and compliance and coronary blood flow during reperfusion of ischemic myocardium (table 2).

An abundant presence for the necessary substrates of XO mediated superoxide generation can be easily conceived. The first substrate, hypoxanthine, has been shown to accumulate in ischemic myocardium as a result of ATP catabolism [50]. During ischemia, ATP is progressively dephosphorylated to adenosine diphosphate (ADP) and finally, to adenosine monophosphate (AMP). The AMP is catabolized to adenosine, inosine, and finally to hypoxanthine. The remaining substrate for the XO activity, molecular oxygen, is supplied during reperfusion of the ischemic heart. This model, as originally proposed by McCord, is shown in figure 3.

A number of investigators supported this hypothesis based on their observations of allopurinol-mediated reduction of reperfusion injury [51–54]. Allopurinol, an analog of hypoxanthine, is a competitive inhibitor of XO. Alloxanthine, or oxypurinol, the metabolite of allopurinol formed by the action of XO, is a potent noncompetitive inhibitor of the enzyme. The formation of this compound, together with its long persistence in tissues, is primarily responsible for much of the pharmacological activity of allopurinol. Recently, the beneficial action of allopurinol to salvage ischemic-reperfused myocardium has been challenged by Reimer and Jennings [55]. In addition, most of the mammalian hearts, with the exception of rat heart, have been found to contain negligible or no XO activity at all [56, 57].

Recently, xanthine dehydrogenase, XO, hypoxanthine and xanthine were assayed from the biopsies obtained from control, ischemic, and reperfused pig hearts. XO activity was not detected in any of the heart biopsies [58]. Although measurable quantities of hypoxanthine were found in the biopsies, only a negligible amount of xanthine was detected. Thus, these results contradict the hypothesis that free radicals may be produced via XO reactions in pig heart. To the contrary, these results support the observations by Reimer and Jennings [55] that XO activity may not contribute to the myocardial ischemic-reperfusion injury. However, contrary to their results, these studies indicated significant protection of heart from reperfusion injury. This discrepancy may be explained due to the species variation, as well as protocol of the experiment such as duration of ischemia, mode of ischemic insult, and dose of allopurinol.

Studies were also conducted to examine the effects of allopurinol and its primary metabolite oxypurinol on the salvage of myocardial ischemic-reperfusion injury when added to the coronary perfusate prior to ischemia. Both oxypurinol and allopurinol were able to preserve the ATP levels at significantly higher values compared to control [16, 58]. CP levels were decreased significantly during ischemia, but their values reached supranormal levels during reperfusion in the treated groups. Cardiac contractility and compliance were also better restored by allopurinol and oxypurinol. LVDP and $LV_{max}dp/dt$ were significantly higher in the treated groups. Thus these results support the previous reports of myocardial protection by XO inhibitors [51–54].

The left ventricular tissue biopsies, when examined at 77 °K using EPR, showed multiple EPR signals (fig. 4) at g = 1.985, 2.000, and 2.012. These EPR signals were significantly amplified in ischemic and reperfused heart, compared to those in perischemic controls. Both oxypurinol and allopurinol were able to reduce these signals in ischemic and reperfused tissue, suggesting free radical scavenging abilities of these XO inhibitors. This study, however, did not identify the EPR spectra generated in heart

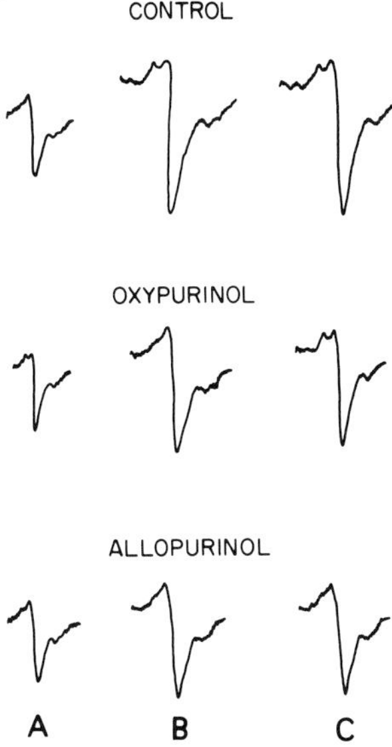

Fig. 4. EPR spectra of (A) nonischemic control, (B) ischemic, and (C) reperfused left ventricular tissue biopsies. Samples were analyzed at 77 °K, microwave power 200 μW, microwave frequency of 9.5 GHz, and modulation amplitude 2.5 G.

biopsies; it showed only enhancement of free radical signals during ischemia and reperfusion, and their reductions by oxypurinol and allopurinol.

In order to identify the free radical scavenged by allopurinol and oxypurinol, oxygen-derived free radicals and hydroxyl radicals were generated by conventional techniques. Freshly isolated PMN were activated with FMLP or ZAS to generate free radicals. Alcide (a gift from Dr. Robert Kross, Alcide Corporation) has been found to generate ClO_2^- and HClO simultaneously. Scavenging actions of oxypurinol and allopurinol were studied by placing these XO inhibitors with the free-radical generating systems. Both EPR and Luminometer were used for the detection of free radicals. Figure 5 and table 3 show the results, which indicate that both oxypurinol and allopurinol (0.1 mM) can scavenge free radicals, but allopurinol seems to be a better scavenger. At a lower concentration, these XO inhibitors do not function as free radical scavengers.

Recent work by Peterson et al. [59] also suggested that allopurinol might function as a free radical scavenger and act as an electron-transfer agent from ferrous iron to ferric cytochrome c. These results support this

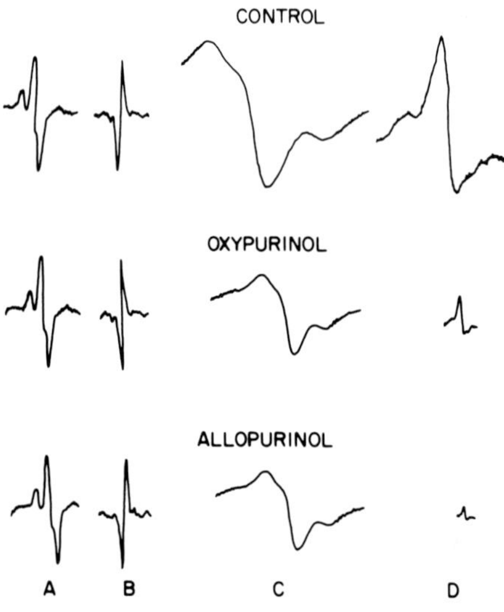

Fig. 5. EPR spectra of (A) oxygen radical generated by the action of KO_2 and DMSO, (B) hydroxyl radical generated from the reaction of Fe^+, DTPA, H_2O_2 and KTBA, (C) PMN activated with FMLP, and (D) chlorine dioxide radical and HOCL generated from Alcide. Samples were analyzed at 77 °K, microwave power 200 μW, microwave frequency 10 GHz, and modulation amplitude 2.5 G.

Table 3. Effects of oxypurinol and allopurinol on various free radical generating systems

Free radical generating systems	Radicals	% inhibition of chemiluminescent response	
		oxypurinol	allopurinol
KO_2 + DMSO	$O_2^{\cdot}$	0	0
Fe^+ + DTPA + H_2O_2 + KTBA	$OH^{\cdot}$	0	0
PMNs + FMLP	$O_2^{\cdot}$ + HOCL	50	50–60
PMNs + ZAS	$O_2^{\cdot}$ + HOCL	50–60	60–70
Alcide	$CLO_2^{\cdot}$ + HOCL	>80	>90

DMSO = Dimethyl sulfoxide; DTPA = diethylenetriamine-pentaacetic acid; KTBA = 2-keto-4-thiomethylbutyric acid; FMLP = n-formyl-methionyl-leucyl-phenylalanine; ZAS = zymosan-activated serum.

earlier hypothesis and further confirm that at higher concentration ($> 10\ \mu M$) these XO inhibitors can scavenge free radicals.

The results of this study clearly indicate that both oxypurinol and allopurinol salvage myocardial functions during ischemia and reperfusion by scavenging free radicals, and not by inhibiting XO. Furthermore, a relatively higher concentration of these XO inhititors must be used, because at lower doses they do not function as scavengers of free radicals.

Mitochondrial Respiration

The generation of H_2O_2 by mitochondrial tissue is an established physiologic event under aerobic conditions [60]. One of the members of respiratory chain, flavoproteins, is known to be responsible for the mitochondrial generation of H_2O_2. The earliest recognized events at a subcellular level under hyperbaric conditions is the inhibition of reversed electron transfer in mitochondria and subsequent formation of oxygen-derived free radicals [61].

The mitochondrial respiratory chain produces superoxide radicals (O_2^-) at two sites at the flavoprotein NADH dehydrogenase and at the ubiquinone-cytochrome c region, likely by autooxidation of ubisemiquinone [62].

In 1971, Loschen et al. [63] first reported that mitochondrial membranes are capable of generating O_2^- in the presence of succinate and antimycin. Four years later, Boveris and Cadenas [64] and Dionisi et al. [65] independently showed that, under the same conditions, O_2^- can be considered as a stoichometric precursor of mitochondrial H_2O_2. Subsequently, ubisemiquinone was shown to be the principal univalent reductant of oxygen in mitochondrial membranes [66]. These findings were further confirmed when Trumpower and Simmons [67] could produce O_2^- by isolated succinate-cytochrome c reductase.

It should be kept in mind, however, that most of the molecular oxygen consumed during mitochondrial respiration is reduced with four electrons to water at the level of cytochrome oxidase. Only a small amount (approximately 5%) of oxygen is reduced univalently to superoxide, which ultimately produces H_2O_2 after dismutation of these radicals.

Catabolic Products of ATP

As previously mentioned, hypoxanthine, a xanthine derivative, is a substrate for XO-induced superoxide anion generation [62, 68–70]. Hypoxanthine has been shown to accumulate in the ischemic myocardium as a result of ATP catabolism [71–74]. Researchers have emphasized the preservation of high-energy phosphate compounds during open-heart surgery [75–78], which is an established parameter for myocardial preservation

[79–82]. However, a direct relation between cardiac function and ATP level is obscured by the compartmentalization of ATP, as shown by Gudbarjarnason et al. [83]. When dephosphorylation of ATP is triggered by a loss of energy balance induced by myocardial ischemia, the cell loses metabolic products of ATP in the form of nucleosides, and the regeneration of ATP becomes limited because of the relatively slow rate of de novo synthesis. During reperfusion, oxidative phosphorylation in combination with myokinase activity results in redistribution of adenine nucleotides reflected in the value of the 'adenylate energy charge', more commonly known as 'energy charge'. This index, defined by $(ATP + 1/2\ ADP)/(ATP + ADP + AMP)$, has been suggested previously as a key factor in the regulation of cellular metabolism [84].

Once nucleosides are generated in the ischemic heart, both hypoxanthine and xanthine can react with XO in the presence of molecular oxygen as discussed earlier (fig. 2). Although the presence of hypoxanthine in most of the mammalian hearts are confirmed, because of the absence of XO, the validity of this hypothetic model remains questionable. However, recent studies demonstrated that the scavenger of superoxide radical superoxide dismutase (SOD), along with catalase, significantly improved myocardial recovery during reperfusion by enhancing rephosphorylation steps [50].

Molecular Oxygen

The observation that reperfusion injury occurs immediately following the onset of reoxygenation suggests the role of molecular oxygen in reperfusion injury [85–87]. Reperfusion of ischemic myocardium is naturally associated with rapid reoxygenation. Molecular oxygen is an essential substrate for the hypoxanthine-XO reaction to produce superoxide radicals. The scheme of the reaction as described by McCord [88] has been diagrammed in figure 3. The role of molecular oxygen in reperfusion injury was suggested by a study in which moderating the rate of reperfusion prevented the development of hemorrhagic infarcts and reduced the high incidence of virulent ventricular arrhythmias associated with reperfusion [89, 90]. In the later study [90] a flow-limiting stenosis was produced, upon which a 30 min complete occlusion was superimposed, followed by reperfusion. The peak flow during reperfusion was induced by the flow-limiting stenosis. Under these circumstances, the authors found a significant inverse relation between survival during reperfusion and peak rate at which blood flow returned to ischemic myocardium. Ventricular fibrillation developed in all cases where peak reperfusion flows exceeded 160% of the baseline flow, but occurred only in 36% of cases whose flow was less than this value.

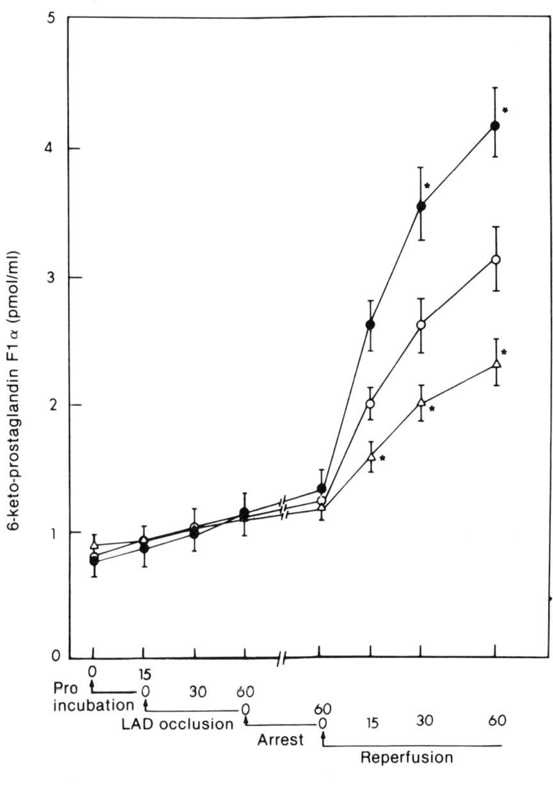

Fig. 6. Time course of change in the level of 6-keto-PGF$_{1\alpha}$ in the perfusate. $\bigcirc$ = Control group (n = 7); $\triangle$ = mepacrine-treated group (n = 7); $\bullet$ = SOD and catalase-treated group (n = 7), where n reflects separate preparations. Each symbol represents mean $\pm$ SEM. *p > 0.05, compared with control group.

Arachidonic Acid Degradation Products

The accumulation of arachidonic acid in the release of prostaglandins have been domonstrated during reperfusion. Further, it has been shown that oxygen-derived free radicals are generated during the conversion of prostaglandin H$_2$ [91]. It has also been shown that free radical scavengers are able to stimulate the degradation of arachidonic acid and enhance the conversion of prostaglandin G$_2$ into prostaglandin H$_2$ [92], suggesting that generated oxygen radicals control the feedback mechanism and deactivate cyclooxygenase enzymes. The absence of such oxygen radicals increase the

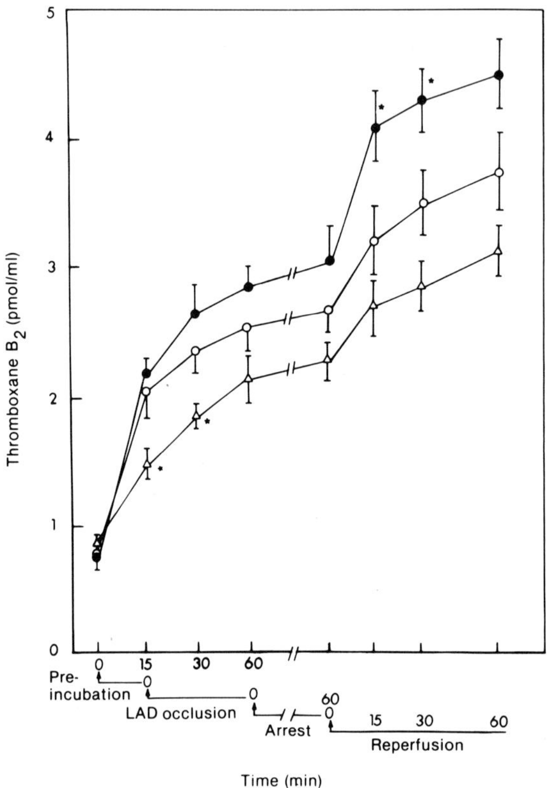

Fig. 7. Time course of change in the level of thromboxane B$_2$ in the perfusate. $\bigcirc$ = Control group (n = 7); $\triangle$ = mepacrine-treated group (n = 7); $\bullet$ = SOD and catalase-treated group (n = 7), where n reflects separate preparations. Each symbol represents mean $\pm$ SEM. *p < 0.05 compared with control group.

activity of cyclooxygenase enzymes by preventing such a feedback mechanism. Recent studies employing mepacrine, a phospholipase inhibitor, and SOD plus catalase, free radical scavengers, support this feedback mechanism as a feature of the naturally protective enzyme systems in the myocardium [39].

A significant amount of 6-keto-PGF$_{1\alpha}$ and thromboxane B$_2$ appeared during reperfusion of ischemic myocardium corresponding with the loss of membrane phospholipids in control animals (fig. 6, 7). Mepacrine protected the depletion of membrane phospholipids and inhibited the products of arachidonate metabolism. SOD and catalase, on the other hand, enhanced the formation of 6-keto-PGF$_{1\alpha}$ and thromboxane B$_2$. While there is no direct evidence to support the mechanism of enhanced prostaglandin synthesis by free radical scavengers, the following assumption can be made. The inhibition of nonenzymatic oxidation of arachidonic acid may stimulate the conversion of free (nonesterified and nonoxidized) arachidonic acid toward prostaglandin synthesis through the cyclooxygenase pathway since a remarkable increase in malonaldehyde formation and its inhibition by SOD and catalase has been documented in this experimental model [8–11]. It has also been shown that MK 447 (2-aminomethyl-4-t-butyl-6-iodophenol) scavenges oxygen free radicals, accelerating the conversion of prostaglandin G_2 to prostaglandin H_2, simultaneously enhancing the production of prostaglandin E_2, thromboxane A_2 and PGI_2 [91]. Thus, it is reasonable to speculate that SOD and catalase, by scavenging oxygen free radicals, promoted the cyclooxygenase pathway and enhanced PGI_2 generation despite the relative decrease in mobilization of arachidonic acid resulting from preservation of membrane phospholipids. On the other hand, inhibition of the appearance of 6-keto-$PGI_{1\alpha}$, by mepacrine is presumably due to the reduction in the availability of free arachidonic acid as a result of inhibition of phospholipases, although inhibition of cyclooxygenase by this compound cannot be excluded [93].

Mechanism of Free Radical Injury

Autooxidation (Nonenzymatic Degredation)
The only defense against the free radicals is the naturally occurring anitoxidative enzymes such as SOD, catalase, or glutathione peroxidase. The activities of these enzymes are known to be depressed during ischemia, limiting the ability of the cell to inactivate the free radical moiety [94, 95]. Further, under ischemic conditions, there is a decreased oxygen concentration and, thereby, a large increase in reducing equivalents persists [70, 96]. The terminal component of the respiratory chain is inhibited, resulting in the reduction of NAD to NADH. With reperfusion and a burst of oxygen, molecular oxygen dissolved in the myocardial cellular lipid matrix is instantly reduced by the respiratory chain carriers [97, 98]. Under these conditions the univalent reduction of oxygen is most likely to occur, thereby favoring the production of superoxide anions and free radicals.

Free radicals readily react with unsaturated fatty acids of the phospholipid membranes [99]. Among the major phospholipids in the cardiac membrane, phosphatidylcholine (PC), phosphatidylethanolamine (PE), and phosphatidylinositol (PI) all contain unsaturated fatty acids [100]. Interaction of free radicals with the phospholipid membrane is likely to cause lipid peroxidation by autooxidation of these unsaturated fatty acids. This could indeed generate new free radicals in a chain reaction which leads to further cell damage during reperfusion and the 'reperfusion injury' [1–17, 101, 102].

Besides causing oxidation of the unsaturated fatty acids in membrane phospholipids, free radicals also react with the accumulated unsaturated fatty acids such as arachidonic acid which forms during myocardial ischemia [13, 36]. It is important to recognize the fact that lipid peroxidation is a free radical-mediated process, and that polyunsaturated fatty acid substrates do not necessarily have to be in the free carboxylate form but can also undergo peroxidation while esterified as phospholipids [103, 104]. This suggests the following hypothesis for lipid peroxidation through the free radical-mediated reperfusion injury:

$$R^\circ + \begin{array}{c} PL \\ \longrightarrow \\ ARA \end{array} \begin{array}{c} RH + PL^\circ \\ \\ \longrightarrow RH + ARA^\circ \end{array} \tag{i}$$

$$\begin{array}{c} PL^\circ \\ \\ ARA^\circ \end{array} \xrightarrow{\;O_2\;} \begin{array}{c} PLO_2 \\ \\ ARA_2^\circ \end{array} \tag{ii}$$

$$\begin{array}{cc} PLO_2^\circ & PLH \\ & + \\ ARAO_2^\circ & ARAH \end{array} \longrightarrow \begin{array}{c} PLO_H + PL^\circ \\ \\ ARAO_2H + ARA \end{array} \tag{iii}$$

$$\begin{array}{cc} PLO_2^\circ & PLO_2H \\ & + \quad \text{metal ions} \\ ARAO_2^\circ & ARAO_2H \end{array} \tag{iv}$$

$$\longrightarrow \begin{array}{l} \text{alkanes} \\ \text{alkanals} \\ \text{alkenals} \\ \text{4-hydroxyalkenals} \end{array}$$

where: R° = a primary reactive free radical in lipid peroxidation; PL = phospholipids; ARA = arachidonic acid.

The foregoing formulae suggest that free radicals, generated during reperfusion of ischemic myocardium, interact with the unsaturated fatty acids and phospholipids to initiate a complex series of reactions (reactions i–iv) that collectively are known as lipid peroxidation. Reactions i and ii can be followed by estimating the fatty acid loss or the oxygen uptake respectively; the formation of PL° and ARA° free radicals are accompanied by bond rearrangement that result in a dienebond character, whereas reaction iii can be monitored by estimating lipid hydroperoxide formation. The final reaction represents a group of complex degradative steps catalyzed by metal ions which ultimately lead to the formation of malonaldehyde. This compound can be monitored by estimating thiobarbituric acid-reactive products [68, 69, 105–107].

Any autooxidation process described above involves three steps: (1) initiation, (2) propagation, and (3) termination. The generally accepted mechanism for autooxidation of an unsaturated fatty acid such as linoleic acid is given below:

$$\text{Initiation: Linoleic acid} \xrightarrow{\text{Initiator}} 5, \text{pentadione radical}$$

$$\text{Propagation: } R^\circ + O_2 \longrightarrow ROO^\circ \longrightarrow ROOH$$

$$\text{Termination: } RO_2^\cdot + {}^\cdot O_2R \longrightarrow \text{nonradical products}$$

Inhibitors of the lipid peroxidation process presumably interfere with the autooxidation steps in the initiation or propagation steps. It is, therefore, common practice to inhibit any autooxidation process by blocking the propagation of the peroxyl radical group with an antioxidant such as α-tocopherol. This phenolic antioxidant can transfer H_2 to peroxy radicals producing a tocopheryl radical which then scavenges a second peroxyl radical.

Phospholipid Degradation by Enzymatic Pathway

The formation of lipid hydroperoxides and malondialdehyde by the enzymatic pathway is quite different from the nonenzymatic schemata previously described. The enzymes phospholipase and cyclooxygenase play a key role in producing phospholipid degradation enzymatically. Phospholipids can be hydrolyzed as a consequence of lipase activation. The phospholipases are known to be activated as a result of calcium influx into the cell, which occurs during ischemia and reperfusion [37, 108–114]. Activation of phospholipase A and C by calcium is well documented. At least three possible enzymatic pathways are known to mediate the release of arachidonic acid from the membrane phospholipids: (a) PI-specific phospholipase C pathway [115]; (b) glycerol lipase pathway [116]; and (c) phosphatidic acid-specific phospholipase A_2 pathway [117]. Since PI of mammalian heart is almost exclusively in the 1-stearoyl-2-arachidonyl form, the PI-specific phospholipase C pathway allows the specific release of arachidonate. Following formation of diglyceride by phospholipase C, arachidonate is either released by di- or monoglyceride lipase, or the

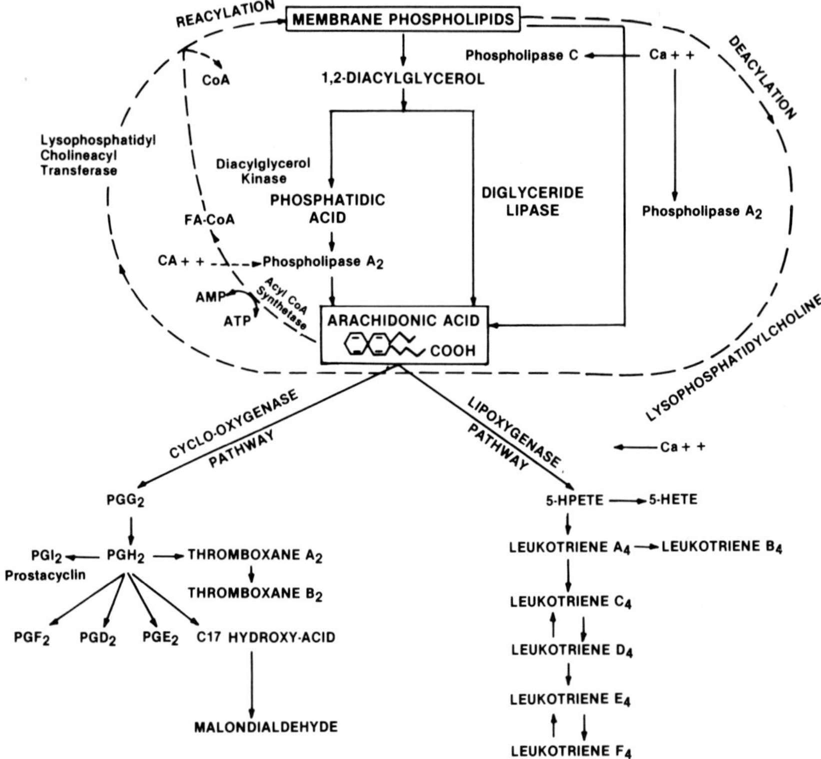

Fig. 8. Enzymatic pathway for the degradation of membrane phospholipids and arachidonic acid.

diglyceride is phosphorylated to phosphatidic acid. Arachidonic acid may then be released from phosphatidic acid by the addition of phosphatidic acid-specific phospholipase A_2. In addition, calcium-stimulated phospholipase A_2 also attacks PC, PE, and PI directly to form arachidonic acid. The enzymatic pathway for the degradation of phospholipids and arachidonic acid is shown in figure 8.

In normal states, arachidonic acid is either reacylated to phospholipids or further degraded via cyclooxygenase or lipooxygenase pathways [118, 119]. But in the presence of ischemia, recent studies suggest a defective reacylation pathway [36, 37] and thereby accumulation of arachidonic acid [36] has been shown to occur. Associated with this is the increased activity of the cyclooxygenase pathway with accumulation of prostaglandins

[121, 122] and the potential for increased formation of oxygen-derived free radicals during the conversion of PGG_2 into PGH_2 [91, 92]. Following the formation of PGH_2, it is broken down enzymatically into the stable compounds of PGE_2, PGF_2, PGD_2, 17-carbon hydroxy acid, 12-hydroxy-5, 8, 10-heptadecatrienoic acid (HHT), and malondialdehyde (fig. 8). PGH_2 is also converted enzymatically into PGI_2 and thromboxanes.

The lipoxygenase pathway has not been adequately explored in heart muscle. However, this enzyme is present in both platelets and PMN [123]. It is possible that lipoxygenase present in the platelets and white cells can be activated as the result of ischemia, and the arachidonic acid accumulated in the myocardium can form leukotrienes through the interaction of this arachidonic acid and the lipoxygenase enzymatic activity of the blood elements. As previously mentioned, PMN tend to accumulate in ischemic myocardial tissue [124–126].

A Comparative Study on Free Radical-Mediated Phospholipid Degradation by Autooxidation and Enzymatic Degradation

It is clear that both pathways of degradation, enzymatic and nonenzymatic, lead to the formation of thiobarbituric acid (TBA)-reactive products (malonaldehyde, the end product of lipid peroxidation). Our laboratory measured the formation of TBA-reactive products during the reperfusion of ischemic myocardium in the presence of a phospholipase inhibitor, mepacrine (one group of animals), and scavengers of free radicals, SOD and catalase (a second group of animals) (fig. 9). Mepacrine-inhibited phospholipid breakdown by enzymatic pathways, and hence the quantity of TBA-reactive material formed in the presence of this lipase inhibitor, was by autooxidation of unsaturated fatty acids and phospholipids alone. The difference in the quantity of TBA-reactive material formed in the presence of this lipase inhibitor was by autooxidation of unsaturated fatty acids and phospholipids alone. The difference in the quantity of TBA-reactive material between control and mepacrine-treated groups is the result of the inhibition of the enzymatic pathway of phospholipid degradation. SOD plus catalase, on the other hand, scavenged the free radicals formed during the reperfusion, and the TBA-reactive material formed is solely by the enzymatic pathway alone. Comparison of these three groups, control, free radical scavengers, and phospholipase inhibition, indicates that about 60% of the malonaldehyde is formed by autooxidation and 40% by enzymatic degradation, suggesting that autooxidation is the major pathway for lipid peroxidation during the free radical attack of reperfusion injury.

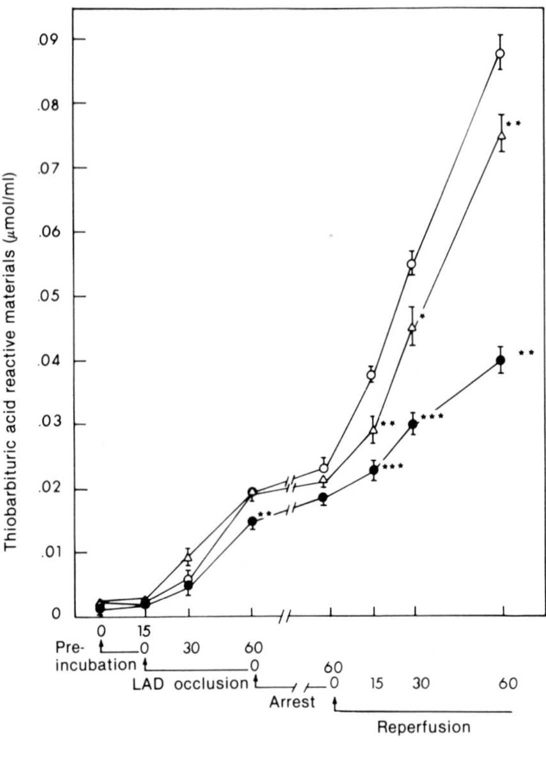

Fig. 9. Time course of change in the level of thiobarbituric acid reactive materials in the perfusate. $\bigcirc$ = Control group (n = 7); $\triangle$ = mepacrine-treated group (n = 7); $\bullet$ = SOD and catalase-treated group (n = 7), where n reflects separate preparations. Each symbol represents mean $\pm$ SEM. *p < 0.05, **p < 0.01, ***p < 0.001 compared with control group.

Interconversion of Oxygen-Derived Free Radicals

Once the O_2^- formed in the system, it rapidly undergoes a series of interconversions leading to the formation of other harmful species as shown in figure 10.

The following represents the sequence of possible biochemical reactions that can occur during the reperfusion of ischemic myocardium:

(i) $O_2^- + O_2^- + 2H^+ \rightarrow H_2O_2 + O_2$
(ii) $H_2O_2 + Fe^2 \rightarrow Fe^3 + {}^\cdot OH + OH^-$
(iii) $O_2^- + H_2O_2 \rightarrow {}^1O_2 + OH^- + {}^\cdot OH$
(iv) ${}^\cdot OH + O_2^\cdot \rightarrow {}^1O_2 + OH^-$

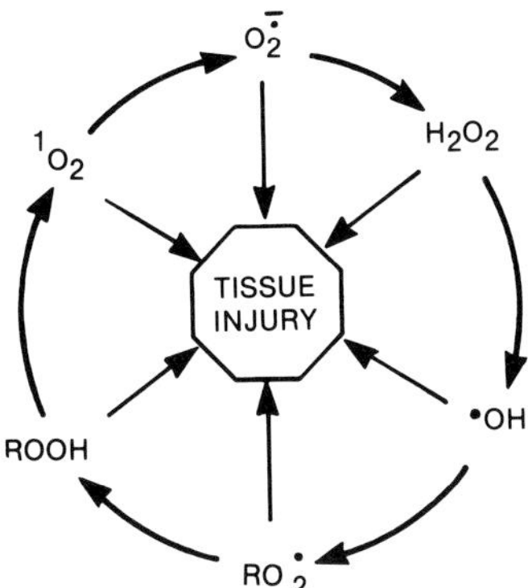

Fig. 10. Tissue injury resulting from superoxide radicals and their interconverted products.

•OH and ^{1}O$_2$ ultimately form peroxy radicals (RO$_2^{\bullet}$), alkoxy radicals (RO$^{\bullet}$), hydroperoxides, and peroxides (ROOH) in the system. Interested readers are referred to the reviews by Chadha [127], Singh and Koppenol [128], Rodgers and Powers [129], Slater [130], Porter and Wagner [131], Pryor [132], Halliwell and Gutteridge [133], and Meerson et al. [134].

Biological Defense of Heart against Free Radical Attack

As mentioned earlier, once free radicals are formed in the ischemic-reperfused heart, they react with polyunsaturated fatty acids also generated during ischemia and reperfusion. However, all biological tissues are protected against free radical injury by the natural occurrence of free radical scavengers. Table 4 lists the free radicals that can be generated during the reperfusion of ischemic myocardium and the corresponding scavengers of these radicals. An extensive study regarding the defense system has been previously reported [135].

The principle antioxidant systems, SOD, catalase, and glutathione peroxidase, are known to be significantly inhibited during myocardial ischemia [11, 94, 95]. A significant amount of free fatty acids, particularly arachidonic acid, is accumulated at the same time. Thus, as soon as reperfusion occurs with a burst of free radicals, the myocardium becomes defenseless against them.

To meet the free radical challenge during reperfusion of ischemic heart, scavengers of free radicals can be introduced prior to reperfusion. Indeed, a number of studies have been performed in recent years where beneficial

Table 4. Free radicals and their scavengers in ischemic-reperfused myocardium

Free Radicals	Mechanism of injury	Scavengers
O_2	Enzyme activation and inactivation Fenton reaction	SOD
$^1O_2$	Peroxidation of unsaturated lipids and phospholipids	Vitamin E
OH	Secondary peroxidation	Antioxidants
RO_2	Secondary peroxidation	Antioxidants
ROOH	Peroxidation Fenton-type Reaction	Glutathione peroxidase
H_2O_2	Peroxidation Fenton-type reaction	Glutathione peroxidase Catalase

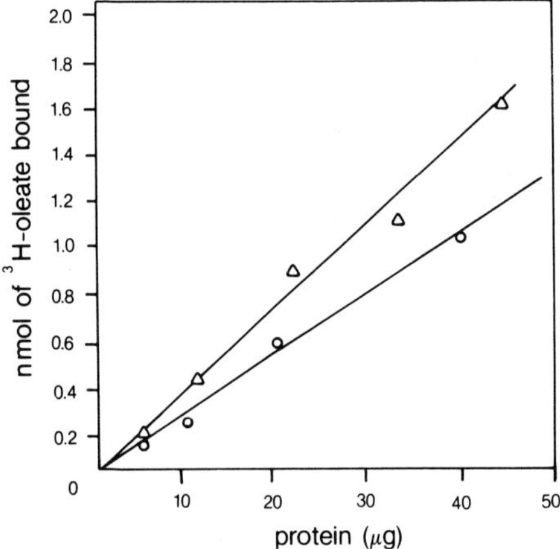

Fig. 11. Binding of [^{3}H]-oleate to delipidated, partially purified FABP from ischemic and nonischemic hearts. The reaction mixture contained 100 mM potassium phosphate (pH 7.4) containing 10 mM KCl, 10 mM [^{3}H]-oleate (K$^+$ salt), and increasing amounts of FABP purified from ischemic ($\triangle$) or nonischemic ($\bigcirc$) hearts, in a total volume of 0.4 ml.

effects of free radical scavengers were observed during reperfusion of ischemic heart [1–16, 39, 136–142].

The substrate for free radicals, polyunsaturated fatty acids such as arachidonic acid, once formed in the ischemic myocardium, cannot combat with any such interventions. These fatty acids cannot be reacylated into the membrane phospholipids because the reacylation system becomes defective during ischemia primarily because of the inactivation of lysophosphatidylcholine acyltransferation [37]. Besides being substrates for the free radicals, these free fatty acids tend to disrupt the membranes by their detergent actions. Recently, a new type of protein has been identified and characterized. This is termed as fatty acid binding protein (FABP) because it exclusively binds the fatty acids and their thioesters [143]. More recently, it has been shown that the partially purified dilipidated FABP isolated from ischemic and nonischemic hearts have similar binding capacities for fatty acids [144]. However, when native cytosols were employed, cytosols of ischemic heart bound 70% less of exogenously added [^{3}H] oleate as compared to nonischemic hearts, suggesting that FABP of ischemic heart might have bound more endogenously fatty acids and their derivatives (fig. 11). Thus, FABP may play a role in the reduction of the concentration of free fatty acids and their metabolites during ischemia.

Detection of Activated Oxygen Species

Direct

Direct measurements of free radicals often require sophisticated instruments, extreme caution, and complicated techniques. Reactions of free radicals with a substrate most often proceed with rates which are only controlled by the diffusion of the reactants, i.e. with absolute rate constants for a bimolecular process in the order of $10^9–10^{10}\ M^{-1}\,\mathrm{s}^{-1}$. This, in turn, yields half-lives for these reactions in the nanosecond to millisecond time range whenever the concentrations of the reaction species exceed micromolar concentrations. These times are considerably shorter than those detectable by conventional or even rapid-mixing techniques [145]. Possible interconversions of the generated free radicals, as discussed previously, also make the accurate determinations impossible. To date, the only reliable method to measure them is by pulse radiolysis methodology, which is extremely sophisticated and expensive, and experimental technique available to only a few laboratories [146, 147].

Another method of detecting free radicals directly is by the use of the spin trapping technique [148–152]. Unfortunately, the theoretical lower limit of free radical detection by this technique using existing instruments is approximately 10 nM. In most circumstances, however, if hyperfine splitting is to be resolved, the practical limit of detection is about 1 μM. Thus, it is only possible to detect radicals which accumulate to this measurable concentration. Rao et al. [153] were the first to directly identify the free radicals in ischemic heart. Although they did not examine the mechanisms by which evolving myocardial ischemia initiated free radical production, they were able to identify the free radical signals using electron spin resonance (ESR) spectroscopy. In this study, approximately 1 g of myocardial tissue biopsies were immediately frozen in liquid N_2 (-196 °C). 200 mg of frozen powdered tissue was mixed with 1 ml isopentane at -140 °C and transferred through a funnel into a 4-mm quartz tube. Free radical contents were measured with an X-band Varian E9 line ESR spectrometer coupled to an Apple II computer. Free radical intensities were calibrated using a stable spin label, 4-hydroxy-2,2,6,6-tetramethyl piper-idinooxy free radical.

Only recently, other investigators were also able to detect free radical directly in myocardial tissue. These studies were performed with [154] or without [15, 16] the use of spin-trapping compounds. In the later study about 250 mg of myocardial biopsies was powdered at 77 °K and packed into 3-mm (ID) ESR quartz tubes to a height of 2 cm and sealed with isopentane to limit the formation and diffusion of liquid O_2 at the specimen surface. The entire packing was performed in a nitrogen environment. The specimen was measured with an IBM ESR spectrometer at 77 °K (fig. 3, 4).

Recently, high-pressure liquid chromatography (HPLC) has been introduced to identify free radicals [155]. The method of detection depends on spin trapping the free radicals using a spin-trapping agent such as DMPO (5,5-dimethylpyrroline N-oxide) followed by injection of a powder tissue sample onto a HPLC unit attached with an electrochemic detector. Two different columns, a Hibar RT LiChrosorb-RP-18, 10 μM, 250 × 4 mm ID, and an Altex Ultra-sphere 3 μM ODS 75 × 4.6 mm ID have been successfully used. The mobile phase consisted of 0.03 M citric acid, 0.05 M sodium acetate, 0.05 M NaOH, and 0.02 M acetic acid, at pH 5.1. The solvent was filtered through a Millipore filter of 0.45 μm pore size.

Indirect

Indirect measurement of the presence of free radicals is based on the formation of the peroxidative end product, malonaldehyde, during the radical attack. Lipid peroxidation is inevitable evidence of free radical injury and irrespective of pathways of peroxidative degradation, malonaldehyde is the final product of peroxidation, as discussed previously. The presence of malonaldehyde can be detected as the TBA-reactive product by well established methods [12, 13, 68, 69, 105–107]. These methods are very simple, rapid, and require minimal handling. To analyze lipid peroxidation by the positive TBA test, an aliquot of the perfusate sample is mixed with 30% trichloroacetic acid, 5 N hydrochloric acid, and 0.75% TBA. The mixture is placed in boiling water for 15 min, and the red color of TBA-malonaldehyde complex was measured with a spectrophometer using a 535-nm wavelength [156]. The sensitivity of this method can be increased severalfold by using a modified TBA method. In this method, a centrifuged sample is mixed with 0.08 N H_2SO_4 followed by the addition of 0.5 ml of 10% phosphotungstic acid. This is centrifuged and the pellet mixed with 2 ml of 0.08 N H_2SO_4 and 0.3 ml of 10% phosphotungstic acid and recentrifuged. The precipitate obtained this way was resuspended in 4 ml of H_2O and 1.0 ml of TBA reagent. The resulting mixture is put in a water bath at 95 °C for 1 h. This is extracted with 5 ml of n-butanol and fluorescence of butanol layer is read at 553 nm with 515 mm excitation. The HPLC system utilizes a 0.39×30 cm uBondapak C_{18} stainless steel analytical column attached to a 3×22 mm guard column packed with C_{18}/coracil with an eluting solvent as 11% HPLC grade methanol in double distilled water. Interested readers are referred to the Methods of Enzymology, volume 105, for further reading regarding details of the methods of free radical estimation.

References

1 Tanaka J, Rominaga R, Yoshitoshi M, et al: Coenzyme Q_{10}: the prophylactic effect on low cardiac output following cardiac valve replacement. Ann Thorac Surg 1981;33:145–151.

2 Shlafer M, Kane PF, Kirsh MM: Superoxide dismutase plus catalase enhances the efficiency of hypothermic cardioplegia to protect the globally ischemic reperfused heart. J Thorac Cardiovasc Surg 1982;83:830–839.

3 Shlafer M, Kane PF, Wiggins VY: Possible role for cytotoxic oxygen metabolites in the pathogenesis of cardiac ischemic injury. Circulation 1982;66(suppl):85–95.

4 Powell WJ, DiBona DR, Flores J, et al: The protective effect of hyperosomotic mannitol in myocardian ischemia and necrosis. Circulation 1976;54:603–615.

5 Lefer AM, Araki H, Okamatsu S: Beneficial actions of a free radical scavenger in traumatic shock and myocardial ischemia. Circ Shock 1981;8:273–282.

6 Meerson FZ, Kagan VE, Kozlov YP, et al: The role of lipid peroxidation in pathogenesis of ishcemic damage and the antioxidant protection of the heart. Basic Res Cardiol 1982;77:465–485.

7 Somers HM, Jennings RB: Experimental acute myocardial infarction. Histologic and histochemical study of early myocardial infarcts induced by temporary or permanent occlusion of a coronary artery. Lab Invest 1964;13:1491–1503.

8 Otani H, Engelman RM, Rousou JA, et al: Cardiac performance during reperfusion improved by pretreatment with oxygen-free radical scavengers. J Thorac Cardiovasc Surg 1986;91:290–295.

9 Das DK, Dobbs WA, Engelman RM, et al: Oxygen-derived free radicals in the pathogenesis of reperfusion injury. Ann NY Acad Sci 1986;463:274–277.

10 Breyer RH, Otani H, Engelman RM, et al: SOD and catalase pretreatment prior to cardioplegic arrest in the face of regional ischemia. Surg Forum 1985;36:269–271.

11 Das DK, Engleman RM, Rousou JA, et al: Pathophysiology of superoxide radical as potential mediator of ischemic and reperfusion injury in pig heart. Basic Res Cardiol 1986;81:155–166.

12 Das DK, Flansaas D, Engleman RM, et al: Age-related development profiles of the antioxidative defense system and the peroxidative status of pig heart. Biol Neonate 1987;51:156–169.

13 Das DK, Engelman RM, Flansaas D, et al: Development profiles of protective mechanisms of heart against peroxidative injury. Basic Res Cardiol 1987;82: 36–50.

14 Jolly SR, Kane WJ, Bailee MB, et al: Canine myocardial reperfusion injury. Its reduction by the combined administration of superoxide dismutase and catalase. Circ Res 1984;54:277–285.

15 Zweier JL, Flaherty JT, Weisfeldt ML: Direct measurement of free radical generation following reperfusion of ischemic myocardium. Proc Natl Acad Sci USA 1987;84:1404–1407.

16 Das DK, Engelman RM, Clement R, et al: Role of xanthine oxidase inhibitor as free radical scavenger: A novel mechanism of action of allopurinol and oxypurinol in myocardial salvage. Biochem Biophys Res Commun, in press.

17 Rao PS, Luber JM: Quantitation of free radicals produced during myocardial ischemia and reperfusion: Comparision of three techniques. Clin Res 1986;34:337A.

18 Romson JL, Hook BG, Kunkel SL, et al: Reduction of ultimate extent of ischemic myocardial injury by neutrophil depletion in the dog. Circulation 1983;67:1016–1023.

19 Romson JL, Hook RB, Rigot VH, et al: The effect of ibuprofen on accumulation of indium-III-labeled platelets and leukocytes in experimental myocardial infarction. Circulation 1982;66:1002–1011.

20 Hill JH, Ward PA: The phlogistic role of C3 leukotactic fragments in myocardial infarcts of rate. J Exp Med 1971;133:885–900.

21 Ford-Hutchinson AW, Bray MA, Doig MV, et al: Leukotriene B_4, a potent chemokinetic and aggregating substance released from polymorphonuclear leukocytes. Nature 1980;186:264–267.

22 Babior BM: Oxygen-dependent microbial killing by phagocytes. N Engl J Med 1978;198:659–688.

23 Badwey JA, Karnovsky ML: Active oxygen species and the functions of phagocytic leukocytes. Ann Rev Biochem 1980;49:695–726.

24 Klebanoff SJ: Oxygen metabolism and the toxic properties of phagocytes. Ann Intern Med 1980;93:480–489.

25 Rosen H, Klebanoff SJ: Hydroxyl radical-generation by polymorphonuclear leukocytes measured by electron spin resonance spectroscopy. J Clin Invest 1979;64:1725–1729.

26 Green MR, Hill HAO, Okolow-Zublowska MF, et al: The production of hydroxyl and superoxide radicals by stimulated human neutrophils – measurement by EPR spectroscopy. FEBS Lett 1979;100:23–26.

27 Harrison JE, Schultz J: Studies on the chlorinating activity of myeloperoxidase. J Biol Chem 1976;251:1371–1374.

28 Cohen HJ, Chovaniee ME, Davies WA: Activation of the guinea pig granulocyte NAD (P)H-dependent superoxide generating enzyme: localization in a plasma membrane enriched particle and kinetics of activation. Blood 1980;55:355–363.

29 Huennekens FM, Felton SP: in Colowick SP, Kaplan ND (eds): Methods in Enzymology. Academic Press, New York, 1957, vol. III, p 950.

30 Babior BM: The enzymatic basis for O_2^- production by human neutrophils. Can J Physiol Pharmacol 1982;60:1353–1358.

31 Goldstein IM, Roos D, Kaplan HB, et al: Complement and immunoglobulins stimulate superoxide production by human leukocytes independently of phagocytosis. J Clin Invest 1975;56:1155–1163.

32 Johnston RB, Lehmeryer JE: Elaboration of toxic oxygen byproducts by neutrophils in a model of immune complex disease. J Clin Invest 1976;57:836–841.

33 Weening RS, Weaver R, Roos D: Quantitative aspects of the production of superoxide radicals by phagocytizing human granulocytes. J Lab Clin Med 1975;85:245–252.

34 Curnelle JT, Badwey JA, Robinson JM, et al: Studies on the mechanism of superoxide release from human neutrophils stimulated with arachidonate. J Biol Chem 1984;259:11851–11857.

35 Katz AM, Messineo FC: Lipid-membrane interactions and the pathogenesis of ischemic damage in the myocardium. Circ Res 1981;48:1–67.

36 Chien KR, Han A, Sen A, et al: Accumulation of unesterified arachidonic acid in ischemic canine myocardium. Circ Res 1984;54:313–322.

37 Das DK, Engelman RM, Rousou JA, et al: Role of membrane phospholipids in myocardial injury induced by ischemia and reperfusion. Am J Physiol 1986;251:471–479.

38 Chatelain P, Latour JG, Tran D, et al: Myocardial infarcts: relation with extent of injury and effect of reperfusion. Circulation 1987;75:1083–1090.

39 Otani H, Engelman RM, Breyer RH, Enhanced prostaglandin synthesis due to phospholipid breakdown in ischemic-reperfused myocardium. Control of its production by a phospholipase inhibitor or free radical scavengers. J Mol Cell Cardiol 1986;18:953–961.

40 Weksler BB, Marcus AJ, Jaffe EA: Synthesis of prostaglandin I (prostacyclin) by cultured human and bovine endothelial cells. Proc Natl Acad Sci USA 1977;74:3922–3926.

41 Ezra D, Boyd LM, Feuerstein G, et al: Coronary constriction by leukotriene C_4, D_4, and E_4 in the intact pig heart. Am J Cardiol 1983;51:1451–1454.

42 Higgs EA, Higgs SA, Moncada S, Vane JR: Prostaglandins and thromboxanes from fatty acids. Prog Lipid Res 1986; 25: 5–11.

43 Levine JD, Lau W, Kwait G: Leukotriene B_4 produced hyperalagesia that is dependent on polymorphonuclear leukocytes. Science 1984;225:743–745.

44 McCord JM, Fridovich: The reaction of cytochrome C by milk santhine oxidase. J Biol Chem 1968;243:5753–5757.

45 Battelli MG, DellaCorte E, Stirpe F: Xanthine oxidase type D (dehydrogenase) in the intestine and other organs of the rat. Biochem J 1972;126:747–749.

46 DeMartino GN, Kuers K: Two calcium dependent calmodulin-stimulated proteases from rat liver. Fed Proc 1981;40:1738.

47 Rousou JA, Engelman RM, Breyer RH, et al: Myocardial salvage by trifluoperazine, a calcium-calmodulin antagonist. Surg Forum, in press.

48 Higgins AJ, Blackburn KJ: Prevention of reperfusion damage in working rat hearts by calcium antagonists and calmoduling antagonsits. J Mol Cell Cardiol 1984;16:427–438.

49 Otani H, Engelman RM, Rousou JA, et al: Improvement of myocardial function by trifluoperazine, a calmodulin antagonist, during experimental acute coronary artery occlusion and reperfusion. J Thorac Cardiovasc Surg 1989;97:267–274.

50 Das DK, Engelman RM, Otani H, et al: Effect of superoxide dismutase and catalase on myocardial energy metabolism during ischemia and reperfusion. Clin Physiol Biochem 1986;f:187–198.

51 DeWall RA, Vasko KA, Stanley EL, et al: Responses of the ischemic myocardium to allopurinol. Am Heart J 1971;82:362–370.

52 Stewart JR, crute SL, Loughlin V, et al: Allopurinol prevents free radical-mediated myocardial reperfusion injury. Surg Forum 1984;35:325–327.

53 DeWall RA, Vasko KA, Stanley EL, et al: Responses of the ischemic myocardium to allopurinol. Am Heart J 1971;81:362–365.

54 Werns SW, Shea MJ, Mitsos SE, et al: Reduction of the size of infarction by allopurinol in the ischemic reperfusion canine heart. Circulation 1986;73:518–524.

55 Reimer KA, Jennings RB: Failure of the xanthine oxidase inhibitor allopurinol to limit infarct size after ischemia and reperfusion in dogs. Circulation 1985;71:1069–1075.

56 Parks DA, Granger DN: Xanthine oxidase: biochemistry, distribution, and physiology. Acta Physiol Scand 1986;548(suppl): 87–99.

57 Eddy LJ, Sewart JR, Jones HP, et al: Free radical-producing enzyme, xanthine oxidase, is undetectable in human hearts. Am J Physiol 1987;253:H709–H711.

58 Das KD, Engelman RM, Clement R, et al: Myocardial salvage by the free radical scavenging action of xanthine oxidase inhibitors; in Simic M (ed): Oxygen Radicals in Tissue Injury. New York, Plenum Publishing, in press.

59 Peterson DA, Kelly B, Gerraro JM: Allopurinol can act as an electron transfer agent. Biochem Res Commun 1986;137:76–79.

60 Boveris A, Chance B: The mitrochondrial generation of hydrogen peroxide. General properties and effect of hyperbaric oxygen. Biochem J 1973;134:707–716.

61 Chance B, Jamieson D, Coles H: Energy-linked pyridine nucleotide reduction: inhibitory effects of hyperbaric oxygen in vitro and in vivo. Nature 1965;206:257–263.

62 Turrens JF, Boveris A: Generation of superoxide axion by the NADH dehydrogenase of bovine heart mitochondria. Biochem J 1980;191:421–427.

63 Loschen G, Flohe L, Chance B: Respiratory chain linked H_2O_2 production in pigeon heart mitochondria. FEBS Lett 1971;18:261–264.

64 Boveris A, Cadenas E: Mitochondrail production of superoxide anions and its relationship to the antimycin insensitive respiration. FEBS Lett 1975;54:311–314.

65 Dionisi O, Galeotti T, Terranova T, et al: Superoxide radicals and hydrogen peroxide formation in mitochondria from normal and neoplastic tissues. Biochim Biophys Acta 1975;403:292–301.

66 Cadenas E, Boveris A, Ragan CI, et al: Production of superoxide radicals and hydrogen peroxide by NADH-ubiquinone reductase and ubiquinol-cytochrome C reductase from beef heart mitochondria. Arch Biochem Biophys 1977;180:247–248.

67 Trumpower BL, Simmons Z: Diminished inhibition of mitochondrial electron transfer from succinate to cytochrome C by thenoyltrifluoacetone by antimycin. J Biol Chem 1979;254:4608–4616.

68 Das DK, Neogi A: Effects of superoxide anions on the (Na^+K) ATpase system in rat lung. Clin Physiol Biochem 1984;2:32–38.

69 Steinberg H, Greenwald RA, Sciubba J, et al: The effect of oxygen-derived free radicals on pulmonary endothelial cell function in the isolated perfused rat lung. Exp Lun Res 1982;3:163–173.

70 Fridovich I: Quantitative aspects of the production of superoxide anion radical by milk xanthine oxidase. J Biol Chem 1970;245:4053–4057.

71 In vitro. 1. Comparison of high energy phosphate production, utilization, and depletion, and of adenine nucleotide catabolism in total ischemia in vitro vs. severe ischemia in vivo. Circ Res 1981;49:892–899.

72 Jennings RB, Schaper J, Hill ML, et al: Effect of reperfusion late in the phase of reversible ischemic injury. Changes in cell volume, electrolytes, metabolites, and ultrastructure. Circ Res 1985;56:262–270.

73 Gebhard MM, Denkhaus H, Sakai K, et al: Energy metabolism and energy release. J Mol Med 1977;2:271–283.

74 Reimer KA, Jennings RB, Hill ML: Total ischemia in dog hearts in vitro. High energy phosphate depletion and associated defects in energy metabolism, cell volume regulation and sarcolemmal integrityu. Circ Res 1981;49:901–911.

75 Engelman RM, Dobbs WA, Rousou JH, et al: Myocardial high energy phosphate replenishment during ischemic arrest. Aerobic versus anaerobic metabolism. Ann Thorac Surg 1981;33:453–458.

76 Zimmer HG, Trendelenburg C, Kammermeir H, et al: De-novo synthesis of myocardial adenine nucleotides in the rat. Circ Res 1973;32:635–642.

77 Erfeinka M, Veech RL, Wilson DF: Thermodynamic relationship between the oxidation reduction reactions and the ATO synthesis in suspensions of isolated heart mitochondria. Arch Biochem Biophys 1974;160:412–421.

78 Pasqua MK, Spray RL, Pellom GL: Ribose enhanced myocardial recovery following ischemia in the isolated working rat heart. J Thorac Cardiovasc Surg 1982;83:390–398.

79 Hearse DJ, Humphrey SM, Bullock RG: The oxygen paradox and the calcium paradox: two facets of the same problem. J Mol Cell Cardiol 1978;10:641–668.

80 Hearse DJ: Oxygen deprivaton and early myocardial contractile failure: a reassessment of the possible role of adenosine triphosphate. Am J Cardiol 1979;44:1115–1121.

81 Vary TC, Angelakos ET, Schaffer SW: Relationship between adenine nucleotide metabolism and irreversible ischemic tissue damage in isolated perfused rat heart. Circ Res 1979;45:218–225.

82 Reibel DK, Rovetto MJ: Myocardial ATP synthesis and mechanical function following oxygen deficiency. Am J Physiol 1980;234:11620–11624.

83 Gudbarjarnason S, Matties P, Ravens KG: Functional compartmentation of ATP and creatine phosphate in heart muscle. J Mol Cell Cardiol 1970;1:325–339.

84 Atkinson DE: The energy change of the adenylate pool as a regulation parameter. Biochemistry 1968;7:4030–4039.

85 Hearse DJ, Humphrey SM, Nayler WG, et al: Border: Ultrastructural damage associated with reoxygenation of the anoxic myocardium. J Mol Cell Cardiol 1975;7:315–324.

86 Hearse DJ: Reperfusion of the ischemic myocardium. J Mol Cell Cardiol 1977;9:605–616.

87 Ganote CE, Kaltenbach IP: Oxygen induced enzyme release: early events and a proposed mechanism. J Mol Cell Cardiol 1974;11:389–406.

88 McCord JM: Oxygen-derived free radicals in postischemic tissue and injury. N Engl J Med 1985;312:159–163.

89 Lucchesi BR, Burmeister WE, Lomas TE, et al: Ischemic changes in the canine heart as affected by the dimethyl quaternary analog of propranolol UM-272 (SC27761). J Pharmacol Exp Ther 1976;199:310–328.

90 Sheehan RH, Epstein SE: Determinants of arrhythmia death due to coronary spasm: effect of pre-existing coronary artery stenosis on the incidence of reperfusion arrhythmia. Circulation 1983;65:259–264.

91 Kuehl FA, Humes JL, Egan RW, et al: Role of prostaglandin endoperoxide PGG_2 inflammatory process. Nature 1977;265:170–173.

92 Kuehl FA, Humes JL, Ham EA, et al: Inflammation: the role of peroxidase-derived products. Adv Prostaglandin Thromboxane Res 1980;6:77–86.

93 Blackwell GJ, Duncombe WG, Flower RJ, et al: The distribution and metabolism of arachidonic acid in rabbit platelets during aggregation and its modification by drugs. Br J Pharmacol 1977;59:353–366.

94 Guarneri C, Glamigni F, Rossoni-Caldarera C: Glutathione peroxidase activity and release of glutathione from oxygen-deficient perfused rat heart. Biochem Biophys Res Commun 1979;89:678–684.

95 Guarneri C, Flamigni F, Caldarera CM: Role of oxygen in the cellular damage induced by reoxygenation of hypoxic heart. J Mol Cell Cardiol 1980;12:797–808.

96 Fridovich I: Hypoxia and oxygen toxicity. Adv Neurol 1979;26:255–259.

97   Badwey JA, Karnovsky ML: Active oxygen species and the functions of phagocytic leukocytes. Annu Rev Biochem 1980;49:695–737.

98   Fee JA, Valentine JS: Chemical and physical properties of superoxide; in Michelson AM, McCord JM, Fridovich I. (eds): Superoxide and Superoxide Dismutases. New York, Academic Press, 1977, p 19.

99   Borme J, Smith EE, Hunter FE: The role of fatty acids in mitochondrial changes during liver ischemia. Arch Biochem Biophys 1970;139:425–430.

100 Dustin GJ, Moncada S, Vane JR: Prostaglandins, their intermediates and precursors: cardiovascular actions and regulatory roles in normal and abnormal circulatory systems. Prog Cardiovasc Dis 1979;21:405–430.

101 Hess ML, Manson NH, Okabe EL: Involvement to free radicals in the pathophysiology of ischemic heart disease. Can J Physiol Pharmacol 1982;60:1382–1389.

102 McCallister LP, Daiello DC, Typers GF: Morphometric observations of the effects of normothermic ischemic arrest on dog myocardial ultrastructure. J Mol Cell Cardiol 1978;10:67–80.

103 Porter NA: in Pryor WA (ed): Free Radicals in Biology. New York, Academic Press, 1980, vol. 4, pp 261–294.

104 Flower R: in Berti F, Velo GP (eds): The Prostaglandin System. New York, Plenum Press, 1980, pp. 27–37.

105 Waravdekar VS, Saslane LD: A sensitive colorimetric method for the estimation of Z-deoxy sugars with the use of malonaldehyde-thiobarbituric acid reaction. J Biol Chem 1959;234:1945–1950.

106 Bird RP, Hung SSO, Hadley M: Determination of malonaldehyde in biological materials by high pressure liquid chromatography. Anal Biochem 1983;128:240–244.

107 Ohkawa LW, Ohishi N, Yagi K: Reactions of linoleic acid hydroperoxide with thiobarbituric acid. J Lipid Res 1978; 19:1053–1057.

108 Nakanishi T, Nishioka K, Jarmakani JM: Mechanism of tissue Ca^{2+} gain during reoxygenation after hypoxia in rabbit myocardium. Am J Physiol 1982; :H437–H449.

109 Henry PD, Schuchleib R, David J, et al: Myocardial contracture and accumulation of mitochondrial calcium in ischemic rabbit heart. Am J Physiol 1977;233:H677–H684.

110 Peng KCF, Kane J, Murphy ML, et al: Abnormal mitochondrial exidative phosphorylation of ischemic myocardium reversed by Ca^{2+} chelating agents. J Mol Cell Cardiol 1977;9:897–908.

111 Shen AC, Jennings RB: Myocardial calcium and magnesium in acute ischemic injury. Am J Pathol 1972;67:417–440.

112 Strunk SW, Smith CW, Blumberg JM: Ultrastructural studies on the lesion produced in skeletal muscle fibers by crude type A *Clostridium perfringens* toxin and its purified alpha fraction. Am J Pathol 1967;50:89–107.

113 Sine KI, Douglas AM, Ricchiute NV: Calcium, strontium and barium movements during ischemia and reperfusion in rabbit ventricle. Circ Res 1978;43:712–720.

114 Jennings RB: Calcium ions in ischemia; in Opie LH (ed): Calcium Antagonists and Cardiovascular Disease. New York, Raven Press, 1984, pp 85–95.

115 Weglicki WB: Degradation of phospholipids of myocardial membrane; in Wildentral K (ed): Degradation Processes: Heart and Skeletal Muscle. Amsterdam, Elsevier/North-Holland Biochemical Press, 1980, pp 377–388.

116 Billah MM, Lapetina EG, Cuatrecasas P: Phospholipase A_2 acitvity specific for phosphatidic acid. J Biol Chem 1981;256:5399–5403.

117 Lapetina EG, Cautrecasas P: Stimulation of phosphatidic acid production in platelets precedes the formation of arachidonate and parallels release of serotonin. Biochim Biophys Acta 1979;574:394–402.

118 Sivakoff M, Pure E, Hsuch W, et al: Prostaglandins and the heart. Fed Proc 1979;38:78–82.

119 Daniel L, King L, Waite M: Source of arachidonic acid for prostaglandin synthesis in Madin-Darby canine kidney cells. J Cell Biol 1981;256:12830–12835.

120 Corr PB, Gross RW, Sobel BE: Amphipathic metabolites and membrane dysfuntion in ischemic myocardium. Circ Res 1984;55:135–154.

121 Hsuch W, Needleman P: Sites of lipase activation and prostaglandin synthesis isolated, perfused rabbit hearts and hydronephrotic kidneys. Prostaglandins 1978;16:661–681.

122 Kraemer RM, Folts JD: Release of prostaglandins following temporary occlusion of the coronary artery. Fed Proc 1973;32:459.

123 Blackwell GJ, DunCombe WG, Flower RJ, The distribution and metabolism of arachidonic acid in rabbit platelets during aggregation and its modification by drugs. Br J Pharmacol 1977;59:353–366.

124 Ylikorvala D, Saarela E, Viinikka L: Increased prostacyclin and thromboxane production in man during cardiopulmonary bypass. J Thorac Cardiovasc Surg 1981;82:245–247.

125 Addonizio VP, Smith LH, Strauss JF, et al: Thromboxane synthesis and platelet secretion during cardiopulmonary bypass with bubble oxygenator. J Thorac Cardiovasc Surg 1980;79:91–96.

126 Davies GC, Sobel M, Salzman EW: Elevated plasma fibrinopeptide A and thromboxane B_2 levels during cardiopulmonary bypass. Circulation 1980;61:808–814.

127 Chadha MS: Singlet Molecular Oxygen. Proc Symp Bombay, Bhaba Atomic Centre, 1975, pp 1–272.

128 Singh A, Koppenol WH: Introduction: interconversion of singlet oxygen and related species. Photochem Photobiol 1978;28:429–433.

129 Rodgers MAJ, Powers EL: Oxygen and Oxy-Radicals in Chemistry and Biology. New York, Academic Press, pp 1–793.

130 Slater TF: Free radical mechanisms in tissue injury. Biochem. J 1984;222:1–15.

131 Porter NA, Wagner CR: Phospholipid autooxidation. Adv Free Radical Biol Med 1986;2:283–323.

132 Pryor WA: Free radicals in autooxidation and aging; in Armstrong D, Sohal RS, Cutler RG, Slater TF (eds): Free Radicals in Molecular Biology, Aging and Disease. New York, Raven Press, 1984.

133 Halliwell B, Gutteridge JMC: Oxygen toxicity, oxygen radicals, transition metals and disease. Biochem J 1984;219:1–14.

134 Meerson FZ, Kagan VE, Kozlov Y et al: The role of lipid peroxidation in pathogenesis of ischemic damage and the antioxidation protection of the heart. Basic Res Cardiol 1982;77:465–485.

135 Fridovich I: The biology of superoxide and of superoxide dismutase – in brief; in (Rodgets MA, Pwers EL (eds): Oxygen and Oxy-Radicals in Chemistry and Biology. New York, Academic Press, pp 197–204.

136 Aust SD, White BC: Iron chelator prevents tissue injury following ischemia. Adv Free Radical Biol Med 1985;1:1–17.

137 White BC, Krause GS, Aust SD, et al: Postischemic tissue injury by iron-mediated free radical lipid peroxidation. Ann Emerg Med 1985;14:804–809.

138 Casale AS, Bolling SF, Rui JA, et al: Progression and resolution of myocardial reflow injury. J Surg Res 1984;37:94–99.

139 Magovern GT Jr, Bolling SF, Casale AS, et al: The mechanism of mannitol in reducing ischemic injury: hyperosmolarity or hydroxyl scavengers. Circulation 1984;70:191–195.

140 Casale AS, Bulkley GB, Bulkely BH, et al: Oxygen-free radical scavengers protect the arrested, globally ischemic heart upon reperfusion. Surg Forum 1983;34:313–316.

141 Chambers DE, Parks DA, Patterson G, et al: Santine oxidase as a source of free radical damage in myocardial ischemia. J Mol Cell Cardiol 1985;17:145–152.

142 Svingen BA, Buege JA, O'Neal FO, et al: The mechanisms of NADPH-dependent lipid peroxidation: the propagation of lipid peroxidation. J Biol Chem 1979;245:5892–5899.

143 Glatz JFC, Paulussen RJA, Veerkamp JH: Fatty acid binding protein from heart. Chem Phys Lipids 1985;38:115–129.

144 Jones R, Prasad R, Engelman R, et al: Effect of ischemia on the fatty acid binding abilities of myocardial fatty acid binding proteins. J Mol Cell Cardiol, in press.

145 Baxendale JH, Busi F: The study of fast processes and transient species by electron pulse radiolysis. NATO Advanced Study Institute Series. Dordrecht, Reidel, 1982.

146 Ebert M, Keene JP, Swallow AJ, et al: Pulse Radiolysis. Cambridge, MIT Press, 1969.

147 Matheson MS, Dorfman LM: Pulse Radiolysis. Cambridge, MIT Press, 1969.

148 Finkelstein E, Rosen GM, Rauckman EJ: Spin trapping of superoxide and hydroxyl radical: practical aspects. Arch Biochem Biophys 1980;200:1–16.

149 Kalyanaraman B, Mottley PC, Mason RP: A direct electron spin resonance and spin-trapping investigation of peroxyl free radical formation by hematin/hydroperoxide systems. J Biol Chem 1983;258:3855–3858.

150 Mason RP, Chignell CF: Free radicals in pharamocology and toxicology-selected topics. Pharmacol Rev 1981;33:189–211.

151 Infold KU: in Free Radicals. New York, Wiley, 1973, vol 1, p 37.

152 Poyer JL, Floyd RA, McCay PB, et al: Spin-trapping of the trichloromethyl radical produced by enzymatic NADPA oxidation in the presence of carbon tetrachloride or bronotrichloromethane. Biochim Biophys Acta 1978;539:402–409.

153 Rao PS, Cohen MV, Mueller HS: Production of free radicals and liquid peroxides in early experimental myocardial ischemia. J Mol Cell Cardiol 1983;15:713–716.

154 Kramer JH, Arroyo CM, Dickens BF, et al: Spin-trapping evidence that graded myocardial ischemia alters post-ischemic superoxide production. Free Radicals Biol Med 1987;3:153–159.

155 Floyd RA, Lewis CA, Wong PK: High-pressure liquid chromatography – electrochemical detection of oxygen free radicals in Packer L (ed): Methods in Enzymology. New York, Academic Press, 1984, vol 105, pp 237–321.

156 Yagi K: Assay for serum lipid peroxide level and its clinical significance; in Yagi K (ed): Lipid Peroxides in Biology and Medicine. New York, Academic Press, 1982, p 223.

Dipak K. Das, PhD, University of Connecticut School of Medicine,
Farmington, CT 06032 (USA)

5 Oxy-Radical Sources, Scavenger Systems and Membrane Damage in Cancer Cells

Tommaso Galeotti, Silvia Borrello, Lanfranco Masotti

Oxygen radicals and related free radical species are generated in the cell as by-products of normal metabolism. For some of these radicals the tissue steady-state concentration has been calculated as well as their lifetime [1–4]. Small molecules and proteins located in the cytosol and membranes or present in extracellular fluids (glutathione, ascorbate, vitamin E, β-carotene, ubiquinone, uric acid, coeruloplasmin and transferrin) act as quenchers of the reactive oxygen species. Moreover, there are enzymes – superoxide dismutase (SOD), glutathione peroxidase and catalase – which are able to scavenge such species as well as the products of lipid metabolism [5]. Further protection to the cell is provided by other enzymatic systems involved in redox reactions (DT-diaphorase, glutathione reductase, methionine sulfoxide reductase, NADH semidehydroascorbate reductase) and repair processes (DNA repair enzymes). Intracellular concentration of oxy-radicals may reach cytotoxic levels (reactions with proteins, DNA, lipids and carbohydrates) as a consequence of an unbalanced prooxidant/antioxidant ratio either in certain prooxidant states, such as hyperoxia, chemical or physical therapy, ischemia, inflammation, blood disorders, or following a decrease in the antioxidant defenses. The tumor cell seems to represent an example where one such condition of oxygen toxicity is accomplished, mainly owing to a deficiency of oxy-radical scavenging enzymes.

This article is meant to be an overview of the membrane sources of reactive oxygen species, of the enzymatic and nonenzymatic defenses and of the membrane lipid damages that are thought to be produced by an oxidative stress, in cancer cells. At the end a unifying hypothesis will be proposed.

1. Subcellular Organelles as Sources of Reactive Oxygen Species

Redox reactions occurring during electron transport in subcellular organelles or membrane networks are responsible for the generation of superoxide anions (O_2^-), hydrogen peroxide (H_2O_2) and hydroxyl radicals (OH$^{\cdot}$). Thus mitochondria, peroxisomes, nuclei, the endoplasmic reticulum and plasmalemmas are sites of oxy-radical formation under physiological conditions [5]. The univalent reduction of oxygen leads to the generation of O_2^-. H_2O_2 is formed either by divalent reduction of dioxygen or monovalent reduction followed by spontaneous or enzymatically catalyzed dismutation [6]. In turn, H_2O_2 may react with O_2^- or reduced iron to give the highly reactive $OH^·$ in the Haber-Weiss [7] and Fenton [8] reactions, respectively. In vivo $OH^·$ formation would occur only via an iron-catalyzed Haber-Weiss reaction [9].

1.1. Mitochondria

Loschen et al. [10] first showed that mitochondrial membranes produce superoxide radicals. Subsequent studies established that the two major sites of O_2^- production are the ubiquinone-cytochrome b region, probably by autoxidation of ubisemiquinone [11–13], and the flavoprotein NADH dehydrogenase [14, 15]. These organelles have also been recognized as a source of intracellular hydrogen peroxide [16–18]. In order to establish the molecular mechanism of H_2O_2 generation, studies have been carried out both in intact mitochondria and submitochondrial particles (SMP). The latter produce higher amounts of O_2^- than the former, as well as small amounts of H_2O_2, due to the unavoidable contamination by SOD. Conversely, in the intact organelle the measurable steady-state concentration of O_2^- is smaller than in SMP due to the fact that the radicals are converted into hydrogen peroxide with a ratio of 2 [19, 20]. Consistently, the addition of SOD to SMP raises the H_2O_2 production to values similar to those of intact mitochondria (fig. 1). The value of this ratio and the behaviour of SMP in the presence of SOD are a good indication that H_2O_2 production in mitochondria is derived from dismutation of O_2^- [10, 11, 19, 20, 22, 23].

Tumor mitochondria also produce O_2^- at rates equal to or higher than those of normal mitochondria. However, they do not produce H_2O_2 (fig. 1). The finding that tumor mitochondria lack SOD [20] provides strong evidence that the precursor of most, if not all, mitochondrial H_2O_2 is O_2^-.

Normal mitochondria have also been reported to produce hydroxyl radicals [24], probably through the reaction of H_2O_2 with ubisemiquinone radicals ($UQ^· + H_2O_2 \rightarrow UQ + OH^· + OH^-$) [25, 26].

1.2. Endoplasmic Reticulum

Another intracellular source of oxy-radicals is the endoplasmic reticulum, as demonstrated by measurements performed on microsomes. This fraction produces superoxide [27–32] and hydroxyl radicals [33, 34] as well as hydrogen peroxide [31, 35, 36]. NADPH-cytochrome c (P-450) reductase and cytochrome P-450 have been proposed to be the sites of O_2^- and H_2O_2 generation. P-450 involvement in H_2O_2 production has been recently

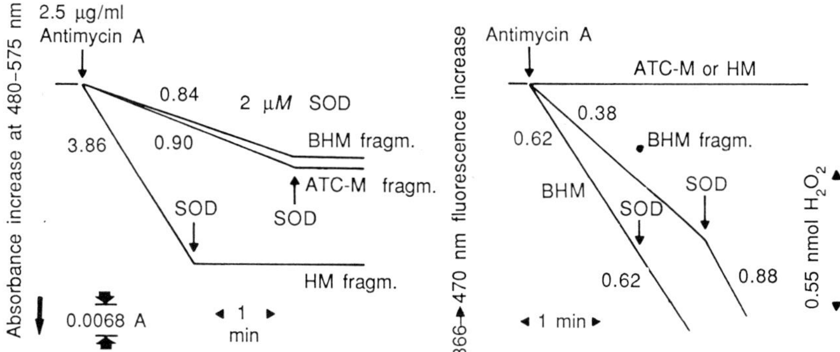

Fig. 1. Production of superoxide radicals (left) and hydrogen peroxide (right) by intact and fragmented mitochondria. The number along the traces indicates nmol O_2^- or H_2O_2 per min^{-1} per mg $protein^{-1}$. The reaction at pH 7.8, in the presence of excess epinephrine in an O_2-saturated medium, involves 1.16 O_2^- radicals per adrenochrome formed [21]. The measurements of O_2^- production shown in this and the following figures were carried out at pH 7.5 in order to avoid autoxidation of epinephrine. BHM = Beef heart mitochondria; ATC-M = Ehrlich ascites tumor cell mitochondria; HM = Morris hepatoma 3924A mitochondria. Adapted from Dionisi et al. [20].

demonstrated also in vivo [37]. Two pathways of oxygen reduction have been proposed: the divalent reduction directly to H_2O_2 and the univalent reduction to O_2^- with subsequent dismutation to H_2O_2. Rat liver microsomes generate $OH^·$ both in the presence [34] and absence [33] of cytochrome P-450 substrate, and require NADPH. The suggested mechanism of $OH^·$ formation is the same as in mitochondria [38].

As shown in table 1, tumor microsomal membranes, produce lower amounts of O_2^- and H_2O_2, owing to the partial deficiency of electron carriers. In fast-growing hepatomas, where cytochrome P-450 is absent, the exclusive source of oxy-radicals seems to be the flavoprotein [39].

The reliability and sensitivity of the adrenochrome assay of O_2^- radicals [41] have been questioned [42, 43]. Yet, the data presented in figure 2 show that most of the adrenochrome formed at pH 7.5, where no autoxidation occurs, is SOD-sensitive and therefore indicative of the presence of O_2^-. However the adrenochrome test cannot be used for a quantitative assay of superoxide radicals since intermediates [42] of the reaction also form this anion, thus resulting in an overestimation of the quantity of superoxide generated. On the other hand, as far as microsomal membranes are concerned [44], the acetylated [45] or succinoylated [46] cytochrome c methods underestimate the rate of O_2^- formation, since both cytochrome

Table 1. O_2^--Dependent epinephrine cooxidation, hydrogen peroxide production and the content of electron carriers in microsomal membranes from rat liver and hepatomas

	Rat liver	Hepatoma 44	Hepatoma 3924A	Novikoff ascites hepatoma
Adrenochrome nmol min^{-1} mg protein^{-1}	24.8 ± 1.7 (13)[a]	10.0 ± 1.1[b] (4)	3.4 ± 0.4[b] (4)	4.7[b] (2)
Hydrogen peroxide nmol mg protein^{-1}	54.4 ± 2.4 (4)[a]	23.3[c] (2)	3.0[c] (2)	NA[e]
NADPH-cytochrome c (P-450) reductase nmol min^{-1} mg protein^{-1}	83.1 ± 3.4 (4)[c]	42.4 ± 1.1[b] (5)	17.8 ± 1.2[b] (5)	25.1[b] (2)
Cytochrome b_5 nmol mg protein^{-1}	0.32 ± 0.02 (5)[a]	0.15 ± 0.01[b] (3)	0.008 ± 0.001[b] (3)	ND[d]
Cytochrome P-450 nmol mg protein^{-1}	0.56 ± 0.06 (6)[a]	0.14 ± 0.02[b] (3)	ND[d]	ND[d]

The values are means ± SE (number of observations).
[a]From ref. [31].
[b]From ref. [39].
[c]From ref. [40].
[d]Not detectable.
[e]Not available.

and SOD cannot penetrate the microsomal membrane. As the adrenochrome values reported in table 1 are meant to compare control and tumor membranes, we still believe that they retain validity.

1.3. Nuclei

The nuclear membrane is another source of oxy-radicals in the cell. It should be stressed that the data so far available have been obtained from nuclear preparations of tumor cells, i.e. Ehrlich [47] and hepatoma 22a [48] ascites cells. The reactive oxygen species experimentally detected were O_2^- and H_2O_2 (fig. 3). Indeed experiments performed with scavengers of OH· and singlet oxygen ($^1O_2$) tended to exclude that these species were formed during nuclear electron transport. At least part of the H_2O_2 found in nuclei could derive from dismutation of O_2^- catalyzed by SOD, present in the fraction [47]. Nuclei are devoid of cytochrome P-450. However, they contain another autooxidizable b-type pigment (peaks at 559, 531 and 428 nm) which is involved as an electron acceptor in NADPH oxidation [49] and might be a source of O_2^- radicals. NADPH-cytochrome c reductase is supposed to be, as in microsomes, the second O_2^- generator [47]. Nuclear membranes from hepatoma 22a ascites cells possess an enzymatic

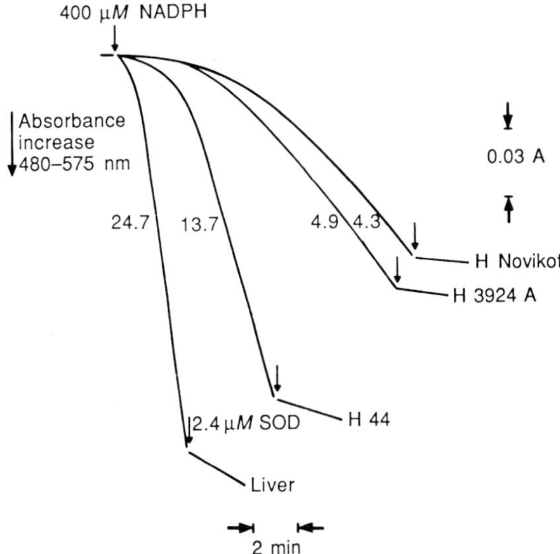

Fig. 2. Cooxidation of epinephrine to adrenochrome in microsomes isolated from rat liver and various hepatomas. The numbers along the traces indicate nmol adrenochrome formed per min^{-1} per mg $protein^{-1}$. H44 = Morris hepatoma 44; H3924A = Morris hepatoma 3924A; H Novikoff = Novikoff ascites hepatoma. Modified from ref. [39].

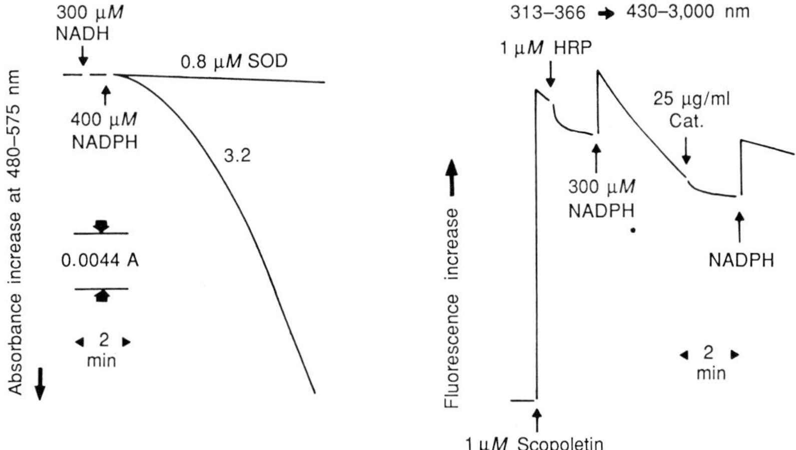

Fig. 3. Production of superoxide radicals as adrenochrome formation (left) and hydrogen peroxide (right) in nuclei from Ehrlich ascites tumor cells. The number along the absorbance trace represents nmol adrenochrome formed per min^{-1} per mg $protein^{-1}$. HRP = Horseradish peroxidase; CAT = catalase. Adapted from ref. [47].

O_2^--generating system that can use both NADH and NADPH as electron donors and seems to involve cytochrome b_5 as the autooxidizable component [48].

2. Antioxidant Defenses

This section will be devoted to a brief outline of the enzymatic systems that the normal cell utilizes for protection against the damaging action of reactive oxygen species, and of the corresponding systems of which the tumor cell is endowed. Data will also be presented on the nonenzymatic free radical scavengers, with particular emphasis on the most important membrane-bound antioxidant, i.e. α-tocopherol.

2.1. Enzymes

In the normal cell several enzymes contribute to maintaining the steady-state concentration of oxy-radicals at a low level. Of these, three have been extensively studied and demonstrated to be of paramount importance: SOD and the two hydroperoxidases, catalase and glutathione peroxidase; some others, which only recently have been either discovered, e.g. methionine sulfoxide reductase [50–52], or considered as free radical scavengers, e.g. DT-diaphorase [53–55], are deemed to play a secondary role.

The cytosolic SOD content was reported by Bozzi et al. [56] to be lowered in fast-growing ascites tumors, such as Ehrlich ascites cells, rabdomyosarcoma MC1-A, Yoshida and Novikoff hepatomas, and by Peskin et al. [57, 58] in other fast-growing ascites and solid tumors of mice (Lewis lung carcinoma, sarcoma 180, adenocarcinoma 755 and Ehrlich ascites cells) and rats (Walker 256 carcinosarcoma, Zajdela hepatoma and hepatoma 27). Subsequent work was mainly addressed at finding a correlation between the tumor growth rate and the enzyme cytosolic content. For this purpose the Morris hepatomas, characterized by different growth rates and degree of differentiation, were preferentially employed [59]. With regard to the cytosolic SOD activity, the data available in the literature [60–62] are presented in figure 4. The enzyme levels were examined in a large spectrum of hepatomas, varying by about 10-fold in growth rate. It can be seen that there is a consistent diminution of the enzyme content with increasing tumor growth rate.

Early studies showed that mitochondrial SOD is absent in SV-40 transformed embryonic cells [63]. It was later observed in fast-growing experimental tumors of the mouse and the rat, namely Ehrlich ascites cells and Morris hepatoma 3924A, that mitochondria are virtually devoid of the

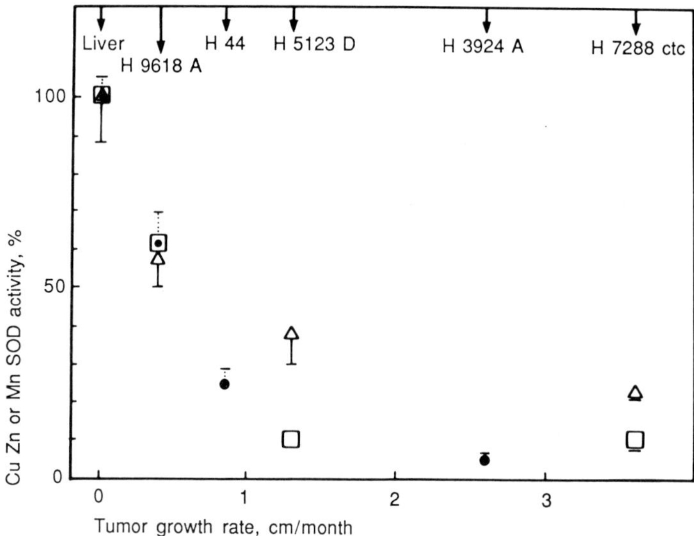

Fig. 4. Percent activity of Cu-Zn or Mn SOD vs tumor growth rate in Morris hepatomas. □ = Mn SOD; △ = Cu-Zn SOD from ref. [61]; ● = Cu-Zn SOD from ref. [60, 62]. The values are given as means ± SE (T from ref. [61]; ⫶ from ref. [60, 62]) of 3–6 experiments.

enzyme [20]. Experiments were subsequently extended to hepatomas with different growth rate, whose results are also reported in figure 4. Again the pattern observed for the cytosolic enzyme is repeated. From the analysis of these data one even more important consideration can be made: poorly differentiated and fast-growing tumors nearly lack compartmental defenses against superoxide radicals, with the exception of nuclei where SOD activity is present, although probably not sufficient for a complete protection from O_2^- [47].

In malignant cells, such as L1210, S91 melanoma and Yoshida sarcoma, Hochstein and Cohen [64] showed a marked detoxication of H_2O_2 by glutathione peroxidase. This observation was confirmed in Ehrlich ascites cells by Hosoda and Nakamura [65]. On the other hand, in other rapidly growing tumors cytosolic glutathione peroxidase was found to be drastically diminished [56, 58]. A more detailed analysis of the GSH peroxidases, the selenoenzyme and the non-selenoenzyme, was carried out in hepatomas again selected for their increasing degree of deviation and growth rate [62]. A different behaviour of the two enzymes was observed: while Se GSH peroxidase activity decreased consistently with increasing proliferation rate of the tumor, the non-Se enzyme did not follow this pattern. In the slow-growing Morris hepatoma 9618A the activity of the latter was in fact found to be even higher than in liver, it was slightly lower than the control in hepatoma 44, with intermediate growth rate, and virtually undetectable in the fast-growing hepatoma 3924A. The pattern of the non-Se GSH peroxidase activity in hepatomas 9618A and 44 has been related to what occurs in states of selenium deficiency [62], where this activity may become essential in counterbalancing the loss of the Se-dependent enzyme [66, 67].

In analogy to what has been found for SOD, glutathione peroxidase is markedly diminished in mitochondria from the fast-growing hepatoma 27 [58].

Catalase, the third important enzyme in the metabolism of the reactive species of oxygen, can be considered an oxy-radical scavenger, even if H_2O_2 is not a radical species. The enzyme, indeed, contributes to maintaining a low intracellular level of highly reactive oxygen species (e.g. OH·) of which H_2O_2 is a precursor [8]. The number of peroxisomes, the subcellular organelles where catalase is mainly localized, has been known for a long time to be lowered in tumor cells and interestingly to be inversely related to the growth rate of the tumor [68]. Consistently, the catalase activity [68] has been found to follow the same pattern as that described for the two oxy-radical scavenging enzymes considered above.

As previously mentioned, a set of redox enzymes provide reduced intermediates (e.g. NADPH, GSH, hydroquinone) to the detoxifying machinery, thus acting as secondary systems in the oxy-radical scavenging activity of the cell. In tumors data are available on the activity of some of these ancillary enzymes, namely glucose 6-phosphate dehydrogenase, glutathione reductase and NADPH:quinone oxidoreductase (DT-diaphorase). It has been known for a long time that malignant cells display a high activity of the hexose monophosphate pathway [69]. The observation that Ehrlich ascites tumor cells possess strong glutathione reductase activity [65, 70], led to the conclusion that these cells GSH levels regulate the availability of $NADP^+$ for the pentose cycle to function at the optimal rate.

Finally, as for DT-diaphorase, it has been demonstrated that in Ehrlich ascites cells the enzyme, in the presence of vitamin K_3, is active in establishing a shunt of reducing equivalents between the cytosol and the mitochondrial respiratory chain [71–73].

2.2. Nonenzymatic Defenses

In the previous section the data available on glutathione, the most important antioxidant of the cytosolic compartment together with ascorbic acid, have been discussed in connection with the enzymatic activities to which the tripeptide is linked.

β-Carotene (vitamin A) [74] and ubiquinone [75, 76] are recognized to be, under particular conditions, effective antioxidants associated to membranes, the former as a singlet oxygen quencher [77], the latter as a scavenger of O_2^- radicals [76]. However, in microsomal membranes from maximal deviation hepatomas their antioxidant activity has not appeared to be significant [78].

Vitamin E, a group of isomers of tocopherol [79, 80], is considered the most important lipophilic antioxidant [81]. It has been proved to be involved in scavenging a wide variety of free radicals (O_2^-, $^1O_2$, OH·, LOO· and other radical species) [82, 83] and its highly protective action within the lipid bilayer [84] seems to be due to the specific interaction between the side chain and the polyunsaturated fatty acid (PUFA) residues (particularly arachidonic acid) of the phospholipids [85]. The range of variability of vitamin E content, expressed as the ratio of the antioxidant to protein or PUFA, in intracellular membranes from different tissues is rather wide, the highest values being found in hyperoxigenated organs [86]. Similarly, membranes isolated from different tumors exhibit a variable content of the vitamin. In fact, poorly differentiated, fast-growing hepatomas, i.e. Novikoff ascites cells and Morris hepatoma 3924A, have a microsomal vitamin E to PUFA ratio markedly higher than the corresponding liver membranes [78, 87], while in the highly differentiated, slow-growing Morris hepatoma 9618A this ratio is consistently lower [88]. However, it should be pointed out that the higher vitamin E to PUFA ratio in rapidly growing tumors is due to the markedly decreased content in PUFA [60, 78], while the vitamin quantitated on a per mg protein basis is virtually unchanged [87].

3. Membrane Lipid Damage

Considering the complex composition and architecture of biomembranes, the potential targets of free radical attack are their three major components: carbohydrates, proteins and lipids.

The alteration of the sugar residues of glycoproteins and proteoglycans [89, 90] can lead to changes in the recognition functions of the cell. As for proteins, a number of amino acid residues have been shown to be modified by oxidative reactions [91]: methionine [92], histidine [93], cysteine [94–96] and proline [97]. Tryptophan, lysine and tyrosine are also known to undergo oxidative reactions, but the detailed mechanisms of their oxidation in proteins have not yet been clarified. Their chemical modifications can lead to structural ones that in turn will trigger functional alterations. Therefore one may anticipate that membrane enzymes, permeases, receptors, electron carriers, and channels can be subjected to modifications leading to changes in vital functions, such as ion and electron transport, metabolite permeability and signal transduction.

However, the far greater damage, that has also been extensively investigated, is to membrane lipids [98]. As a consequence of free radical reactions, phospholipid and cholesterol undergo chemical changes that deeply alter the static and dynamic properties of the bilayer. Being the membrane a dynamic unit, where all components are strictly interrelated and where lipids do not play a mere supporting role, the oxidative damage to this component concurs in expanding the negative effect of the radical action. Our attention in the following paragraphs will be restricted to the process of lipid peroxidation in the different cellular components.

Data on lipid peroxidation in cancer cells are available from the literature of several years ago and refer to experiments carried out on mitochondrial preparations and homogenates from poorly differentiated tumors [99–104]. It was observed that one feature common to these preparations is a noticeable resistance to the action of peroxidizing agents. Some of these studies were aimed at testing the effect of nonenzymic, such as ascorbic acid, or enzymic, i.e. NADPH, prooxidants in ascites cell [99, 100] and hepatoma D30 mitochondrial membranes [105], respectively. Further research was directed at finding a relationship between tumor mitochondria peroxidizability and the cell proliferating activity [106]. To this end, two hepatomas of the Morris line, 3924A (fast-growing) and 44 (intermediate growth rate), were compared to rat liver. NADPH-dependent lipid peroxidation decreased with increasing growth rate. At the same time, Ehrlich ascites tumor cell mitochondria exhibited low lipid peroxidation values, comparable to those of the hepatoma 3924A.

Among the subcellular membranes, microsomes have been the subject of extensive investigation, being a useful model to study the mechanisms and effects on membranes of free radical damage [5]. Player et al. [107] employed several rapidly growing hepatomas (D23, D30 and D192A) and found that NADPH-dependent lipid peroxidation was entirely absent in microsomal fractions. Bartoli and Galeotti [106] showed that in microsomes from Ehrlich ascites cell, and Morris hepatomas 3924A and 44, the rate of NADPH-dependent lipid peroxidation was quite similar and much lower than that of liver. By contrast, when exogenous peroxidizing agents (i.e. ascorbate or xanthine plus xanthine oxidase) were employed, the peroxidation rate was still somewhat lower than in liver, but differences between tumors, related to the proliferating activity, were observed. These results preliminarily suggested that different rate-limiting factors contribute to the resistance of microsomal membranes to lipid peroxidation. Specifically, in the NADPH-dependent process, the reaction rate is limited by the availability of both the electron transport system components (table 1) and lipids as substrate, while in the presence of exogenous prooxidants the major limiting factor, at least in the initiation step, are the PUFA. Indeed, in subsequent work [60, 87, 108] it was shown that these hepatomas differ in the lipid composition and content: the higher was the growth rate of the tumor, the lower were the microsomal phospholipid content and the degree of fatty acid unsaturation. More specifically, the highly differentiated Morris hepatoma 9618A had a lipid composition and content very similar to those of hepatocytes [87, 108] and in the presence of superoxide anions, generated by xanthine plus xanthine oxidase, also exhibited a similar peroxidation rate [87] (fig. 5). On the other hand, microsomes from this hepatoma lack from 40 to 60% of the liver cytochrome P-450 content [87, 88] and this explains why in the presence of O_2^- radicals the rate of the initiation reaction does not differ substantially from that of liver microsomes, whereas a slow down of the propagation step does occur when sufficient LOOH accumulate within the membrane [87]. Further evidence that cytochrome P-450 availability is rate-limiting in the peroxidation process was afforded by experiments performed with organic hydroperoxides (e.g. t-butylhydroperoxide) which require the hemoprotein both for the activation of ROOH to initiating radicals (ROO˙, RO˙) and the activation of LOOH to propagating radicals (LOO˙, LO˙). In these experiments, as shown in figure 5, the entire process of lipid peroxidation in microsomes from the hepatoma 9618A was significantly depressed [88]. Evidence for the limiting role of cytochrome P-450 in one or both the steps of lipid peroxidation is also given by experiments performed with microsomes isolated from the hepatoma 9618A of rats treated with the cytochrome P-448 inducer 3-methylcholanthrene. Restoration of normal cytochrome levels caused the almost complete recovery of the microsome peroxidizability [109].

As mentioned in section 2.2, vitamin E plays a key role in protecting membranes from free radical induced peroxidation. It has been suggested [78] that the low response of poorly differentiated tumors to peroxidizing agents could be attributed to the increase in the vitamin content relative to PUFA. However, since vitamin E-deficient hepatoma 3924A microsomes did not show an increased susceptibility to peroxidation with respect to control tumor microsomes, its role in the high resistance of tumor membranes to peroxidation seems to be ruled out [87].

Tumor plasma membranes were also tested for their ability to undergo lipid peroxidation. Comparative experiments were performed in rat liver and hepatoma 3924A plasmalemmas by using xanthine plus xanthine

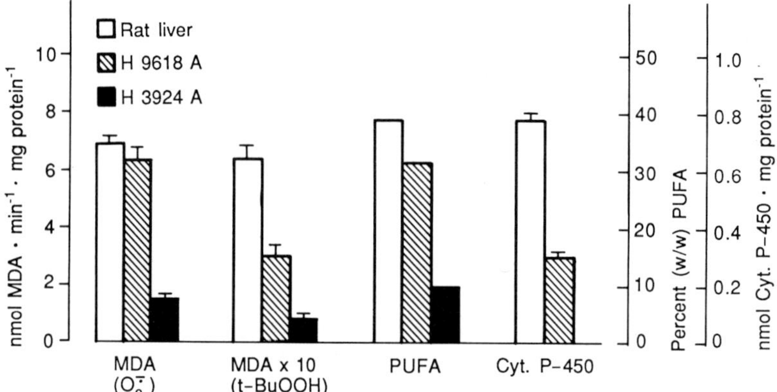

Fig. 5. Superoxide and hydroperoxide-dependent lipid peroxidation of rat liver and hepatoma microsomes as a function of PUFA and cytochrome P-450 content. Superoxide radicals were generated by the xanthine plus xanthine oxidase system. The rates of malondialdehyde (MDA) production are initial velocities calculated from the linear portion of the curve (first 5 min). The real MDA values in the t-BuOOH system are obtained by dividing those reported in the figure by a factor of 10. MDA and cytochrome P-450 values are means $\pm$ SE of 4–10 and 10–11 experiments, respectively. Cytochrome P-450 was undetectable in hepatoma 3924A microsomes. The PUFA content (mean of 3–4 preparations) refer to linoleic and arachidonic acids. Data adapted from Borrello et al. [114].

oxidase as an O_2^- generating system. While the rate of malondialdehyde accumulation was extremly low, LOOH were practically undetectable. Such a high resistance to peroxidation has been attributed to the marked decrease of lipid content and therefore to substrate availability for the radical attack [110].

The picture that emerges from the studies on plasmalemma and some intracellular membranes, aimed at finding a correlation between peroxidizability and degree of deviation of the cancer cell, is that of a cell highly resistant to the peroxidative stress, but that also shows a susceptibility to oxy-radicals inversely related to its proliferating activity. One more cellular parameter to be considered as being at the origin of this behaviour is the architecture of the membrane. Its composition and structural organization can indeed modulate the extent of the peroxidative attack and some of its effects could affect the static and dynamic properties of the membrane, thus leading to some of the misfunctions that characterize the tumor cell. Physical methods, able to monitor events that occur in the nanosecond time scale, are the proper approach to gain insight in this type of molecular phenomena.

Tumor microsomal membranes and plasmalemmas have been studied by fluorescence depolarization measurements of 1,6-diphenyl-1,3,5-hexatriene (DPH) and trimethylammonium diphenylhexatriene (TMA-DPH). For microsomes [111] it has been shown that the bilayer becomes more ordered as the growth rate of the hepatoma investigated increases. Compared to rat liver microsomes, fluidity decreases without varying much in the different hepatomas [111]. Normal plasma membranes, on the other hand, are already highly ordered and do not differ in this respect from plasmalemmas of a fast-growing tumor, such as Morris hepatoma 3924A. Fluidity, instead, decreases noticeably in this hepatoma [110]. Measurements of the order parameter and fluidity carried out on intact cells (Ehrlich ascites tumor cells, lymphocytes from chronic lymphatic and hairy cell leukemias), by employing the fluorescent probe TMA-DPH, confirm this trend [112]. Physical properties quite comparable to those of tumor membranes have been observed by subjecting normal membranes to O_2^--dependent lipid peroxidation [113].

4. Conclusions

The studies carried out in cancer cells in vitro on the sources of oxy-radicals, on the defences employed to gradually transform these in less reactive and less damaging molecular species, on the targets and the ensuing transformation of small molecules or macromolecules, that play particularly significant structural and/or functional roles, have led to discover some interesting features of these cells.

Malignant cells are resistant to peroxidation and such resistance is correlated to the degree of deviation of the tumor: the less differentiated and faster growing is the tumor, the less it is susceptible to the reactive species of oxygen. The enzymatic antioxidant defences are depressed in tumor cell compartments: the higher the growth rate of the tumor, the lower the content of SOD, glutathione peroxidase and catalase. Associated to these features, a rather extensive damage to cellular membranes has also been demonstrated. The composition of the lipid bilayer is changed, particularly in terms of a marked decrease in PUFA residues. Concomitantly, as expected, the static and dynamic properties of the membranes are changed: generally, in solid tumors there are both an increase in order and a decrease in fluidity, as far as intracellular membranes are concerned. Contrariwise, only a decrease in fluidity is observed in plasma membranes. In systemic tumors, measurements of such parameters in intact cells show the same trend for plasmalemmas. These observations are consistent with the decrease of the phospholipid/protein and the increase of the cholesterol/

phospholipid ratios in plasma and intracellular membranes. In any case, the change in the molecular dynamics of the membrane can cause enzymatic proteins or receptors to be more or less exposed on the membrane surface or to be shedded altogether, with major functional consequences.

Important studies that have been carried out on normal cells have shown that lipid peroxidation by-products do indeed inhibit cell division [115]. If one considers the very low production of lipidoxy-radicals exhibited by cancer cells, the hypothesis that the deficiency of these metabolites causes the tumor cells to lose, at least partially, the ability to control their growth rate seems to be reasonably supported. The initial damage to the cell could be represented by the reduction of the antioxidant defenses, to which a higher sensitivity to the action of oxy-radicals would follow. The ensuing transformations, i.e. loss of phospholipids and of polyunsaturated acyl chains would lead in a late stage to a diminished production of peroxidation intermediates capable of exerting some control in cell proliferation.

However, as already mentioned at the beginning of this section, it has to be stressed that all the experimental evidence supporting this working hypothesis has been obtained by studies in vitro. That is to say, whichever will be the effect of the oxy-radical attack on the cell components, and whatever its dynamics, one has until now dealt with cells completely tranformed and malignant. The resistance to peroxidation of these cells would therefore be the 'end-product' of the overall process. If a role then of the peroxidative process in tumor growth has to be unequivocally established, studies on transformed cells during division should constitute the next step of experimentation. This would allow us to establish a correlation between the degree of deviation and the quantity and chemical nature of the peroxidation products, as well as define the role of the latter in interfering and controlling the mechanisms of cell proliferation.

Acknowledgements

This work was supported by grants from MPI 40%, AIRC and CNR, Special Project 'Oncology', Grant N. 86.00423.44 (T.G.) and Grant N. 86.00476.44 (L.M.). We thank Mrs. Giovanna Pelosi and Mr. Claudio Agostini for assistance in the preparation of the manuscript.

References

1   Oshino N, Chance B, Sies H, et al: The role of H_2O_2 generation in perfused rat liver and the reaction of catalase compound I and hydrogen donors. Arch Biochem Biophys 1973;154:117–131.

2   Tyler DD: Polarographic assay and intracellular distribution of superoxide dismutase in rat liver. Biochem J 1975;147:493–504.

3   Boveris A, Cadenas E, Chance B: Ultraweak chemiluminescence: a sensitive assay for oxidative radical reactions. Fed Proc 1981;40:195–198.

4   Pryor WA: Oxy-radicals and related species: their formation, lifetimes and reactions. Ann Rev Physiol 1986;48:657–667.

5   Freeman BA, Crapo JD: Biology of disease: Free radicals and tissue injury. Lab Invest 1982;47:412–426.

6   Fridovich I: Oxygen radicals, hydrogen peroxide and oxygen toxicity; in Pryor WA (ed): Free Radicals in Biology. New York, Academic Press, vol 1, 1976, pp 239–277.

7   Haber F, Weiss JJ: The catalytic decomposition of hydrogen peroxide by iron salts. Proc R Soc Lond Ser A 1934;147:332–351.

8   Fenton HJH: Oxidation of tartaric acid in the presence of iron. J Chem Soc 1984;65:899–910.

9   Halliwell B, Gutteridge JMC: Oxygen toxicity, oxygen radicals, transition metals and disease. Biochem J 1984;219:1–14.

10   Loschen G, Azzi A, Richter C, et al: Superoxide radicals as precursor of mitochondrial hydrogen peroxide. FEBS Lett 1974;42:68–72.

11   Boveris A, Cadenas E, Stoppani AOM: Role of ubiquinone in the mitochondrial generation of hydrogen peroxide. Biochem J 1976;156:435–444.

12   Cadenas E, Boveris A, Ragan CI, et al: Production of superoxide radicals and hydrogen peroxide by NADH-ubiquinone reductase and ubiquinol-cytochrome c reductase from beef heart mitochondria. Arch Biochem Biophys 1977;180:248–257.

13   Turrens JF, Freeman BA, Levitt JG, et al: The effect of hyperoxia on superoxide production by lung submitochondrial particles. Arch Biochem Biophys 1982;217:401–410.

14   Forman HJ, Kennedy J: Dihydroorotate-dependent superoxide production in rat brain and liver. Arch Biochem Biophys 1976;173:219–224.

15   Turrens JF, Boveris A: Generation of superoxide anion by the NADH dehydrogenase of bovine heart mitochondria. Biochem J 1980;191:421–427.

16   Loschen G, Flohé L, Chance B: Respiratory chain linked H_2O_2 production in pigeon heart mitochondria. FEBS Lett 1971;18:261–264.

17   Boveris A, Oshino N, Chance B: The cellular production of hydrogen peroxide. Biochem J 1972;128:617–630.

18   Boveris A, Chance B: The mitochondrial generation of peroxide. General properties and effect of hyperbaric oxygen. Biochem J 1973;134:707–716.

19   Boveris A, Cadenas E: Mitochondrial production of superoxide anions and its relationship to the antimycin insensitive respiration. FEBS Lett 1975;54:311–314.

20   Dionisi O, Galeotti T, Terranova T, et al: Superoxide radicals and hydrogen peroxide formation in mitochondria from normal and neoplastic tissues. Biochim Biophys Acta 1975;403:292–300.

21   Misra HP, Fridovich I: The generation of superoxide radical during the autoxidation of ferredoxins. J Biol Chem 1971;246:6886–6890.

22   Nohl H, Hegner D: Do mitochondria produce oxygen radicals in vivo? Eur J Biochem 1978;82:563–567.

23   Nohl H, Breuninger V, Hegner D: Influence of mitochondrial radical formation on energy-linked respiration. Eur J Biochem 1978;90:385–390.

24   Nohl H, Jordan W, Hegner D: Identification of free hydroxyl radicals in respiring rat heart mitochondria by spin trapping with the nitrone DMPO. FEBS Lett 1981;123:241–244.

25 Nohl H, Jordan W, Hegner D: Mitochondrial formation of OH radicals by an ubisemiquinone-dependent reaction. An alternative pathway of the iron-catalyzed Haber-Weiss cycle. Hoppe-Seylers Z Physiol Chem 1982;363:599–607.

26 Nohl H, Jordan W: The biochemical role of ubiquinone and ubiquinone-derivatives in the generation of hydroxyl radicals from hydrogen peroxide; in Bors W, Saran M, Tait D (eds): Oxygen Radicals in Chemistry and Biology. Berlin, de Gruyter, 1984, pp 155–163.

27 Strobel HW, Coon MJ: Effect of superoxide generation and dismutation on hydroxylation reactions catalyzed by liver microsomal cytochrome P-450. J Biol Chem 1971;246:7826–7829.

28 Aust SD, Roerig DL, Pederson TC: Evidence for superoxide generation by NADPH-cytochrome c reductase of rat liver microsomes. Biochem Biophys Res Commun 1972;47:1133–1137.

29 Estabrook RW, Baron J, Franklin M, et al: Cytochrome P-450 – Panacea or plague; in Shultz J, Cameron BF (eds): The Molecular Basis of Electron Transport. New York, Academic Press 1972, pp 197–230.

30 Mishin V, Pokrovsky A, Lyakovich VV: Interactions of some acceptors with superoxide anion radical formed by the NADPH-specific flavoprotein in rat liver microsomal fraction. Biochem J 1976;154:307–310.

31 Bartoli GM, Galeotti T, Palombini G, et al: Different contribution of rat liver microsomal pigments in the formation of superoxide anions and hydrogen peroxide during development. Arch Biochem Biophys 1977;184:276–281.

32 Estabrook RW, Verringloer J: The oxygen sensing characteristics of microsomal enzymes. Adv Exp Med Biol 1977;78:19–35.

33 Cohen G, Cederbaum AI: Chemical evidence for production of hydroxyl radicals during microsomal electron transfer. Science 1979;204:66–68.

34 Hawco F, Hulett L, O'Brien PJ: Hydroxyl radical formation during P-450 function and benzopyrene hydroxylation; in Coon MJ, Conney AH, Estabrook RW, Gelboin H, Gillette JR, O'Brien PJ (eds): Microsomes, Drug Oxidations and Chemical Carcinogenesis. New York, Academic Press, 1980, vol 1, pp 419–422.

35 Thurman RC, Ley HG, Scholz R: Hepatic microsomal ethanol oxidation, hydrogen peroxide formation and the role of catalase. Eur J Biochem 1972;25:420–430.

36 Hildebrandt AG, Speck M, Roots I: Possible control of hydrogen peroxide production and degradation in microsomes during mixed function oxidation reaction. Biochem Biophys Res Commun 1973;54:968–975.

37 Premereur N, Van den Branden C, Roels F: Cytochrome P-450-dependent H_2O_2 production demonstrated in vivo. Influence of phenobarbital and allylisopropylacetamide. FEBS Lett 1986;199:19–22.

38 Cederbaum AI, Cohen J: Microsomal oxidant radical production and ethanol oxidation. Methods Enzymol 1984;105:516–522.

39 Galeotti T, Bartoli GM, Bartoli S, et al: Superoxide radicals and lipid peroxidation in tumour microsomal membranes; in Bannister WH, Bannister JV (eds): Biological and Clinical Aspects of Superoxide Dismutase. New York, Elsevier/North Holland, 1980, pp 106–117.

40 Bartoli GM, Galeotti T: Unpublished results.

41 Misra HP, Fridovich I: The role of superoxide anion in the autoxidation of epinephrine and a simple assay for superoxide dismutase. J Biol Chem 1972;247:3170–3175.

42 Bors W, Michel C, Saran M, et al: The involvement of oxygen radicals during the autoxidation of adrenaline. Biochim Biophys Acta 1978;540:162–172.

43 Bors W, Saran M, Lengfelder E, et al: Detection of oxygen radicals in biological reactions. Photochem Photobiol 1978;28:629–639.

44 Smith MT, Thor H, Orrenius S: Detection and measurement of drug-induced oxygen radical formation. Methods Enzymol 1984;105:505–510.

45 Azzi A, Montecucco C, Richter C: The use of acetylated ferricytochrome *c* for the detection of superoxide radicals produced in biological membranes. Biochem Biophys Res Commun 1975;65:597–603.

46 Kuthan H, Tsuji H, Graf H, et al: Generation of superoxide as a source of hydrogen peroxide in a reconstituted monooxygenase system. FEBS Lett 1978;91:343–345.

47 Bartoli GM, Galeotti T, Azzi A: Production of superoxide anions and hydrogen peroxide in Ehrlich tumour cell nuclei. Biochim Biophys Acta 1977;497:622–626.

48 Peskin AV, Zbarsky IB, Konstantinov AA: A novel type of superoxide generating system in nuclear membranes from hepatoma 22a ascites cells. FEBS Lett 1980;117:44–48.

49 Bartoli GM, Dani A, Galeotti T, et al: Respiratory activity of Ehrlich ascites tumor cell nuclei. Z Krebsforsch Klin Onkol 1975;83:223–231.

50 Abrams WR, Weinbaum G, Weissbach L, et al: Enzymatic reduction of oxidized alpha-1-proteinase inhibitor restores biological activity. Proc Natl Acad Sci USA 1981;78:7483–7486.

51 Carp H, Janoff A, Abrams WR, et al: Human methionine sulfoxide-peptide reductase, an enzyme capable of reactivating oxidized alpha-1-proteinase inhibitor in vitro. Ann Rev Respir Dis 1983;127:301–305.

52 Morrison HM, Burnett D, Stockley RA: The effect of catalase and methionine-S-oxide reductase on oxidized alpha-1-proteinase inhibitor. Biol Chem Hoppe-Seyler 1986;367:371–378.

53 Lind C, Hochstein P, Ernster L: DT-diaphorase as a quinone reductase: a cellular control device against semiquinone and superoxide radical formation. Arch Biochem Biophys 1982;216:178–185.

54 Thor H, Smith MT, Hartzell P, et al: The metabolism of menadione (2-methyl-1,4-naphtoquinone) by isolated hepatocytes. A study of the implications of oxidative stress in intact cells. J Biol Chem 1982;257:12419–12425.

55 Wefers H, Sies H: Hepatic low-level chemiluminescence during redox cycling of menadione and the menadione-glutathione conjugate: relation to glutathione and NAD(P)H: quinone reductase (DT-diaphorase). Arch Biochem Biophys 1983;224:568–578.

56 Bozzi A, Mavelli I, Finazzi-Agro A, et al: Enzyme defense against reactive oxygen derivatives.II.Erythrocytes and tumour cells. Mol Cell Biochem 1976;10:11–16.

57 Peskin AV, Zbarsky IB, Konstantinov AA: An examination of the superoxide dismutase activity in tumour tissue. Dokl Acad Nauk SSSR 1976;229:751–754.

58 Peskin AV, Koen YM, Zbarsky IB, et al: Superoxide dismutase and glutathione peroxidase activities in tumours. FEBS Lett 1977;78:41–45.

59 Morris HP, Wagner BP: Induction and transplantation of rat hepatomas with different growth rate (including 'minimal deviation' hepatomas); in Busch H (ed): Methods Cancer Res. New York, Academic Press, 1968, vol 4, pp 125-152.

60 Bartoli GM, Bartoli S, Galeotti T, et al: Superoxide dismutase content and microsomal lipid composition of tumors with different growth rates. Biochim Biophys Acta 1980;620:205–211.

61 Bize IB, Oberley LW, Morris HP: Superoxide dismutase and superoxide radical in Morris hepatomas. Cancer Res 1980;40:3686–3693.

62 Bartoli GM, Galeotti T, Borrello S, et al: Loss of defensive enzymes and oxygen-mediated damage of microsomal membranes in liver and hepatomas, in Galeotti T, Cittadini

A, Neri G, Papa S (eds): Membranes in Tumour Growth. Amsterdam, Elsevier, 1982, pp 461–470.

63 Yamanaka NY, Deamer D: Superoxide dismutase activity in WI-38 cell cultures. Effects of age, trypsinization, and SV-40 transformation. Physiol Chem Phys 1974;6:95–106.

64 Hochstein P, Cohen G: The action of hydrogen peroxide on tumor cells. Proc Am Assoc Cancer Res 1962;3:329.

65 Hosoda S, Nakamura W: Role of glutathione in regulation of hexose monophosphate pathway in Ehrlich ascites tumor cells. Biochim Biophys Acta 1970;222:53–64.

66 Lawrence RA, Burk RF: Glutathione peroxidase activity in selenium-deficient rat liver. Biochim Biophys Res Commun 1976;71:952–958.

67 Prohaska JR, Ganther HE: Glutathione peroxidase activity of glutathione-S-transferases purified from rat liver. Biochim Biophys Res Commun 1977;76:437–445.

68 Mochizuki Y, Hruban Z, Morris HP, et al: Microbodies of Morris hepatomas. Cancer Res 1971;31:763–773.

69 Weber G: Carbohydrate metabolism in cancer cells and the molecular correlation concept. Naturwissenshaften 1968;55:418–429.

70 Williams-Ashman HG: Studies on the Ehrlich ascites tumor.II.Oxidation of hexose phosphates. Cancer Res 1953;13:721–725.

71 Dallner G, Ernster L: Abolition of the Crabtree effect in Ehrlich ascites tumor cells. Exp Cell Res 1962;27:368–372.

72 Gordon EE, Ernster L, Dallner G: Intracellular hydrogen transport in Ehrlich ascites tumor cells. Cancer Res 1967;27:1372–1377.

73 Galeotti T, Azzi A, Chance B: The reoxidation of cytoplasmic reducing equivalents in Ehrlich ascites tumor cells. Biochim Biophys Acta 1970;197:11–24.

74 Burton GW, Ingold KU: Beta-carotene: an unusual type of lipid antioxidant. Science 1984;224:569–573.

75 Landi L, Cabrini L, Sechi AM, et al: Antioxidative effect of ubiquinones on mitochondrial membranes. Biochem J 1984;222:463–466.

76 Littarru GP, Lippa S, De Sole P, et al: In vitro effect of different ubiquinones on the scavenging of biologically generated O_2^-. Drugs Exp Clin Res 1985;11:529–532.

77 Kearns DR: Physical and chemical properties of singlet oxygen. Chem Rev 1971;71:395–427.

78 Cheeseman KH, Collins M, Proudfoot K, et al: Studies on lipid peroxidation in normal and tumour tissues. Biochem J 1986;235:507–514.

79 Machlin LJ (ed): Vitamin E: A comprehensive Treatise. New York, Dekker, 1980.

80 Porter R, Whelan J (eds): Biology of Vitamin E, Ciba Fdn Symp 101. London, Pitman, 1983.

81 Burton GW, Joyce A, Ingold KU: Is vitamin E the only lipid-soluble, chain-breaking antioxidant in human blood plasma and erythrocyte membranes? Arch Biochem Biophys 1983;221:281–290.

82 Ozawa T, Hanaki A, Matsumoto S, et al: Electron spin resonance studies of radicals obtained by the reaction of α-tocopherol and its model compounds with superoxide. Biochim Biophys Acta 1978;531:72–78.

83 Nishikimi M, Yamada H, Yagi K: Oxidation by superoxide of tocopherol dispersed in aqueous media with deoxycholate. Biochim Biophys Acta 1980;627:101–108.

84 Cadenas E, Ginsberg M, Rabe U, et al: Evaluation of alpha-tocopherol antioxidant activity in microsomal lipid peroxidation as detected by low-level chemiluminscence. Biochem J 1984;223:755–759.

85 Diplock AT, Lucy JA: The biochemical modes of action of vitamin E and selenium: a hypothesis. FEBS Lett 1973;29:205–210.

86   Kornbrust DJ, Mavis RD: Relative susceptibility of microsomes from lung, heart, liver, kidney, brain and testes to lipid peroxidation: correlation with vitamin E content. Lipids 1980;15:315–322.

87   Borrello S, Minotti G, Palombini G, et al: Superoxide-dependent lipid peroxidation and vitamin E content of microsomes from hepatomas with different growth rates. Arch Biochem Biophys 1985;238:588–595.

88   Minotti G, Borrello S, Palombini G, et al: Cytochrome P-450 deficiency and resistance to t-butyl hydroperoxide of hepatoma microsomal lipid peroxidation. Biochim Biophys Acta 1986;876:220–225.

89   Greenwald RA: Effect of oxygen-derived free radicals on connective tissue macro-molecules; in Bannister WH, Bannister JV (eds): Biological and Clinical Aspects of Superoxide and Superoxide Dismutase. Amsterdam, Elsevier, 1980, pp 160–171.

90   Wong SF, Halliwell B, Richmond R, et al: The role of superoxide and hydroxyl radicals in the degradation of hyaluronic acid induced by metal ions and by ascorbic acid. J Inorg Biochem 1981;14:127–134.

91   Pryor WA: The role of free radical reactions in biological systems; in Pryor WA (ed): Free Radicals in Biology. New York, Academic Press, 1976, vol 1, pp 1–43.

92   Brot N, Weissbach H: Biochemistry and physiological role of methionine sulfoxide residues in proteins. Arch Biochem Biophys 1983;223:271–281.

93   Fucci L, Oliver CN, Coon MJ, et al: Inactivation of key metabolic enzymes by mixed function oxidation reactions: possible implication in protein turnover and aging. Proc Natl Acad Sci USA 1983;80:1521–1525.

94   Brigelius R, Schult E: The role of hepatic $NADP^+$ and glutathione status altered by oxidative stress in the regulation of pentose phosphate cycle; in Rotilio G, Bannister JV (eds): Oxidative Damage and Related Enzymes; Life Chemistry Reports. London, Harwood Academic Publishers, 1984, suppl II, pp 277–283.

95   Gilbert HF: Redox control of enzyme activities by thiol/disulfide exchange. Methods Enzymol 1984;107:330–351.

96   Ziegler DM: Role of reversible oxidation-reduction of enzyme thiols-disulfides in metabolic regulation. Ann Rev Biochem 1985;54:305–329.

97   Wolff SP, Garner A, Dean RT: Free radicals and protein degradation. Trends Biochem Sci 1986;11:27–31.

98   Halliwell B, Gutteridge JMC: Lipid peroxidation: a radical chain reaction; in Halliwell B, Gutteridge JMC (eds): Free Radicals in Biology and Medicine. Oxford, Clarendon Press, 1986, pp 139–189.

99   Thiele EH, Huff JW: Lipid peroxide production and inhibition by tumor mitochondria. Arch Biochem Biophys 1960;88:208–211.

100   Utsumi K, Yamamoto G, Inaba K: Failure of Fe^{2+} induced lipid peroxidation and swelling in the mitochondria isolated from ascites tumor cells. Biochim Biophys Acta 1965;105:368–371.

101   Lash ED: The antioxidant and prooxidant activity in ascites tumors. Arch Biochem Biophys 1966;115:332–336.

102   Fonnesu A, Del Monte U, Olivotto GA: Consumo d'ossigeno e perossidazione lipidica delle frazioni citplasmatiche extramitocondriali di fegato e di epatoma ascite di Yoshida. Lo Sperimentale 1966;116:353–372.

103   Ugazio G, Gabriel L, Burdino E: Ricerche sugli inibitori della perossidazione lipidica presenti nelle cellule dell'epatoma ascite di Yoshida. Boll Soc Ital Biol Sper 1968;44:30–34.

104   Burlakova EB, Molochkina EM, Pal'mina NP: Role of membrane lipid oxidation in control of enzymatic activity in normal and cancer cells. Adv Enzyme Regul 1980;18:163–179.

105 Player TJ, Mills DJ, Horton AA: NADPH-dependent lipid peroxidation in mitochondria from livers of young and old rats and from rat hepatoma D30. Biochem Soc Trans 1977;5:1506–1508.

106 Bartoli GM, Galeotti T: Growth-related lipid peroxidation in tumour microsomal membranes and mitochondria. Biochim Biophys Acta 1979;574:537-541.

107 Player TJ, Mills DJ, Horton AA: Lipid peroxidation of the microsomal fraction and extracted microsomal lipids from the DAB-induced hepatomas. Br J Cancer 1979;39:773–778.

108 Hostetler KY, Zenner BD, Morris HP: Phospholipid content of mitochondrial and microsomal membranes from Morris hepatomas of varying growth rates. Cancer Res 1979;39:2978–2983.

109 Borrello S, Galeotti T, Palombini G, et al: Restoration of hydroperoxide-dependent lipid peroxidation by 3-methycholanthrene induction of cytochrome P-448 in hepatoma microsomes. FEBS Lett 1986;209:305–310.

110 Galeotti T, Borrello S, Palombini G, et al: Lipid peroxidation and fluidity of plasma membranes from rat liver and Morris hepatoma 3924A. FEBS Lett 1984;169:169–173.

111 Masotti L, Cavatorta P, Sartor G, et al: A fluorescence depolarization investigation of membranes from tumour cells with different growth rate; in Galeotti T, Cittadini A, Neri G, Papa S (eds): Membranes in Tumour Growth. Amsterdam, Elsevier Biomedical Press, 1982, pp 39–50.

112 Masotti L, Cavatorta P, Sartor G, et al: Static and dynamic properties of plasma membranes of whole tumour cells; in Galeotti T, Cittadini A, Neri G, Papa S, Smets LA (eds): Cell membranes and Cancer. Amsterdam, Elsevier, 1985, pp 91–96.

113 Masotti L, Cavatorta P, Ferrari MB, et al: O_2^--dependent lipid peroxidation does not affect the molecular order in hepatoma microsomes. FEBS Lett 1986;198:301–306.

114 Borrello S, Minotti G, Galeotti T: Factors influencing O_2^- and t-BuOOH-dependent lipid peroxidation of tumor microsomes; in Rotilio G (ed): Superoxide and Superoxide Dismutase in Chemistry, Biology and Medicine. Amsterdam, Elsevier 1986, pp 432–434.

115 Cornwell DG, Marisaki N: Fatty acid paradoxes in the control of cell proliferation: prostaglandins, lipid peroxides and co-oxidation reactions; in Pryor WA (ed): Free Radicals in Biology. New York, Academic Press, 1984, vol 6, pp 95–148.

Tommaso Galeotti, MD, Institute of General Pathology, Catholic University,
Largo F. Vito 1, I–00168 Roma (Italy)

6 Free Radicals and Skeletal Muscle Disorders

Malcolm J. Jackson

Free radicals play an essential physiological role in normal muscle function, particularly in the mitochondria the movement of electrons (i.e. free radicals by definition) is utilised as a means of energy transfer in the electron transport chain. A less well established and investigated role for free radicals is as a pathogenic agent in various aspects of skeletal muscle pathology. It is likely that skeletal muscle during normal activity suffers from a considerable insult of types likely to initiate free radical-induced damage. Skeletal muscle is the tissue of the body which has the largest requirement for energy during exercise such that there is a vast increase in mitochondrial activity and in oxygen consumption by the muscle. Since many of the radicals proposed to have damaging effects on tissues are oxygen centred, this increase in oxygen throughput provides an increased precursor of these radicals and could predispose the muscle to this type of damage. In addition skeletal muscle unlike other tissues undergoes severe physical stresses and shearing forces during exercise which may induce formation of free radical species.

The aim of this chapter is to discuss the possible roles which free radicals play in various pathological disorders of skeletal muscle. It is important, in consideration of the evidence for a role of free radicals in various pathological processes, to critically consider the methods which have been used to examine these substances in muscle and other tissues, since many different techniques have been utilised and some of these are now recognised to be of dubious value in studies of skeletal muscle. In addition it is felt that a proper evaluation of the possible pathological roles which free radicals could play in skeletal muscle tissue would be incomplete without some discussion of the nature of the damage which free radicals are capable of causing to skeletal muscle. At the present time free radicals are the subject of much interest and many speculative claims are made for the involvement of these species in pathological disorders. In this review only those disorders of skeletal muscle in which it is thought that reasonable evidence for free radical involvement has been presented are discussed and a proposed integrating hypothesis of the role of free radicals in skeletal muscle disorders is presented.

Methods of Assessment of Free Radical Activity in Skeletal Muscle Tissue

In order to demonstrate that decreased free radical activity plays some role in pathological processes in skeletal muscle, it is necessary to be able to provide evidence of their increased activity in appropriate muscle samples. This is a technically difficult procedure since there is no single accepted test for evidence of increased free radical activity in biological materials and muscle tissue presents certain unique problems. Essentially the techniques utilised at the present time are either indirect indices of free radical activity in which the products of reactions between free radicals and biological compounds are monitored. Of these only one, the efflux of volatile hydrocarbons, can be monitored non-invasively, while others can be measured in blood plasma or in biopsy samples of muscle. Direct measurements of free radicals in muscle have been infrequently undertaken using electron spin resonance (ESR) techniques.

Non-Invasive Measurements of Exhaled Hydrocarbons

The release of the hydrocarbons ethane and pentane from peroxidising polyunsaturated fatty acids of the three and six series, respectively, has been utilised as a means of non-invasively monitoring the free radical-mediated peroxidation of fatty acids [1–4]. An increased excretion is thought to be indicative of increased peroxidation within the body, but unfortunately provides no indication of the site of this peroxidation. Both ethane and pentane are present at relatively high levels in the background air and it is necessary for the subject in the experiment (either animals or man) to breathe hydrocarbon-free air prior to the experiment. The use of this technique is undoubtedly of benefit in studies of sedentary animals or man, but the validity of the method in situations where blood flow to tissues varies has been severely questioned [5]. It appears that certain hydrocarbons such as pentane are extremely soluble in tissues (such as skeletal muscle or fat) and that an increase in blood flow to tissues releases considerable amounts of stored hydrocarbons (e.g. pentane) which may not therefore reflect the indigenous rate of lipid peroxidation. This work appears to invalidate the studies of Dillard et al. [1] and Gee and Tappel [4] who studied pentane excretion during exercise in man and animals, respectively, since exercise induces substantial increases in blood flow to muscle with a consequent efflux of pentane from tissue stores. Snider et al. [5] were further able to demonstrate the validity of these criticisms by demonstrating a considerable increase in isopentane excretion during exercise in man. The exhalation of this air pollutant with no known biological origin was increased during exercise by a similar amount to that of pentane.

Measurement of Indirect Indicators of Free Radical Activity in Body Fluids

It is inherent in the chemistry of free radical species that they will interact rapidly with a variety of cellular components and in certain circumstances the nature of the products of these reactions are well described. The interaction of various different radical species with polyunsaturated lipids in the process of lipid peroxidation is one such relatively well understood system, and the products and intermediates of this reaction are those most frequently monitored as indirect indices of free radical activity. Of these the thiobarbituric acid reactivity (or malonaldehyde content) and the level of diene conjugates [6] have been the most frequently studied together with examination of various fluorescent materials which may reflect the incidence of free radical activity [7]. These parameters have been extensively studied in accessible body fluids such as blood plasma from patients with muscular dystrophy [8–10]. Unfortunately measurement of these parameters in blood gives no indication of their source such that a finding of high levels of an indicator may be due to an increase in lipid peroxidation in tissues other than muscle. Alternatively it is possible that peroxidation in any one particular tissue (such as muscle) may not be reflected in the plasma level.

There is also a further possible objection to the use of one of these indicators, the fluorescent pigments, in studies of skeletal muscle tissue. These pigments share many of the characteristics of lipofuscin and may be produced in a similar manner, mainly the interaction of lipid peroxidation products such as malonaldehyde with amino acids and proteins [7]. However, our preliminary studies (unpublished) suggest that free radical damage to muscle homogenates 'in vitro' produces only small amounts of these pigments in comparison to the production of other indicators (e.g. malonaldehyde or diene conjugates) or by comparison with other tissues such as liver.

Several workers have also attempted to examine the activity of the various different protective enzymes and levels of antioxidants in body fluids to obtain some indication of the likelihood of pathological abnormalities of free radical metabolism in muscle [9–13]. The rationale behind these studies being that a low level of protective agents will indicate an increased susceptibility to free radical damage, while an elevated level may occur as a response to an increased oxidative stress. Obviously studies such as these only provide indirect evidence and would need to be supported by further investigations in order to draw firm conclusions.

Measurement of Indirect Indicators of Free Radical Activity in Muscle

The various indices which have been discussed in the biological fluids can also be studied directly in muscle tissue. Adequate amounts of muscle tissue are readily obtained from man using the needle biopsy technique [14]. However, in practice only one indicator, the muscle thiobarbituric acid (TBA) reactive material (or malonaldehyde) has been extensively studied. This has been found to be elevated in muscle from patients with Duchenne muscular dystrophy [13, 15, 16] and in animal muscle following exhaustive exercise [12, 17]. Unfortunately although the results of the measurements of these products in the tissue of interest would appear to be less prone to misinterpretation than measurements in body fluids, these techniques are not without their drawbacks, since Halliwell and Gutteridge [18] have pointed out that in many circumstances damaged tissues may lead to an initiation of lipid peroxidation rather than vice versa. Measurement of elevated levels of lipid peroxidation products in tissues from patients suffering from degenerative muscle disorders may therefore reflect damage to the tissue rather than an increase in pathogenic free radical-mediated processes. In addition malonaldehyde can be an unreliable index of lipid peroxidation in intact biological tissues since it is an intermediate rather than an end product of lipid peroxidation having been shown to be metabolised by intact cells [19]. The production of TBA-reactive material has also been monitored as an index of the susceptibility of muscle to lipid peroxidation in animals with selenium and/or vitamin E deficiency [20, 21]. Similar studies have also been undertaken monitoring the volatile hydrocarbons (e.g. ethane) released from muscle homogenates subjected to an oxidant stress [20].

Direct Measurement of Free Radicals in Muscle Tissue

Direct detection of free radicals can be achieved by ESR spectrometry. This technique has had wide applications in the field of physical chemistry, but has only been relatively sparsely used to examine biological materials. Its use in the study of biological processes has been limited by the high water content of biological materials since water strongly absorbs microwaves. Early investigators attempting to utilise this technique in the study of tissues examined lyophilised material [22]. It later became apparent that signals obtained from dry material frequently resulted from the lyophilisation process or from subsequent changes [23]. Some studies have been undertaken on fresh tissue [12] but a more satisfactory solution to the problem appears to be rapid freezing of the tissue. This should provide ESR spectra of the radicals as they were at the time of freezing [24].

One major free radical signal with a g value 2.0036–2.004 is found in skeletal muscle [12, 24]. Unfortunately this is a very broad signal and provides little information concerning the nature of the radical(s) from which it is derived, but Davies et al. [12] have obtained evidence that the amplitude of this signal is increased with exhaustive exercise. In our studies [24] excess contractile activity was found to induce both damage to the muscle resulting in a loss of intracellular enzymes and an increase in the amplitude of the major free radical signal.

However, a problem which is associated with these studies is that there is no firm relationship established between the ESR signal from intact tissues and oxidative damage to tissues. These problems require further study.

Nature of the Damage to Skeletal Muscle Caused by Free Radical Species

There is little experimental work examining the nature of the damage which increased free radical activity can cause to skeletal muscle. Experimentally rats injected intraperitoneally with large amounts of ferrous salts suffer considerable non-specific damage to the diaphragm and other muscles. This is thought to be due to the production of free radical species by iron-catalysed reactions [25]. In addition isolated sarcoplasmic reticulum exposed to a source of superoxide radicals loses the ability to handle calcium [26], while mitochondria exposed to superoxide radicals may lose the ability to retain calcium [27]. We have also found that isolated muscle exposed to low concentrations of organic peroxides, such a t-butyl hydroperoxide 'leaks' substantial amounts of intracellular constituents to the extracellular medium [28].

In addition it would be expected that free radical mediated damage to the lysosomal membranes would be deleterious to muscle via release of lysosomal enzymes and it should also be noted that free radicals can cause substantial damage to proteins [29] particularly the sulphydryl groups of proteins and since the contractile machinery of skeletal muscle is particularly rich in these then possible damage in this area should not be ignored.

Muscle Disorders in which Free Radicals May Play a Pathogenic Role

Vitamin E and Selenium Deficiency Myopathies in Animals
Vitamin E and selenium deficiency in domestic animals cause a group of disorders which in certain countries result in serious economic loss. The aetiology of the presentation of these disorders is complex involving

vitamin E, selenium, other antioxidants, sulphur-containing amino acids, unsaturated fatty acids and other dietary constituents. The cause of the clinical presentation of the disorder may therefore not necessarily involve an overt deficiency of one factor, but the product of various factors.

Vitamin E and selenium deficiency diseases have deleterious effects on various organs and tissues in different species, such as the liver (swine and rats), brain (rats, poultry, monkeys, man) and heart (swine, cattle, sheep), but the most prominent feature in many commercially important species is a skeletal muscle degeneration, particularly found in young sheep and cattle, but also appearing in swine, rats, mice, rabbits and chickens. This disease has been known by a variety of names of which the most commonly used are white muscle disease or nutritional muscular dystrophy although Bradley and Fell [30] have stated that the description of this disorder as a muscular dystrophy is a misnomer and the disease should be called a nutritional myodegeneration.

The muscle disease typically takes the form of stiffness, weakness, difficulties in rising and tremor. Cardiac involvement is common in young animals and the intercostal muscles may be affected. The gross pathology is of whitish-yellow areas in affected muscles (hence the name white muscle disease). The myodegeneration may preferentially effect type one (red slow oxidative) fibres and in surviving animals there is evidence of substantial regeneration with time. In the acute phase of myodegeneration there is a rise in the activity of muscle-derived enzymes (e.g. creatine kinase and aldolase) in plasma.

The disease in farm animals primarily occurs in animals fed on crops deficient in either selenium or vitamin E. Naturally occurring selenium-deficient soils are present in many parts of the world while there have been various outbreaks of vitamin E deficiency in farm animals whose foodstuffs have been stored in a form which inadvertently induces vitamin E degradation (i.e. storage of cereals with propionic acid as an antifungal agent [31].

One of the most interesting aspects of this disease in cattle has been a marked seasonal variation in the incidence of the disease with the greatest incidence occurring in the spring following 'turnout' after maintenance indoors throughout the winter. This has led to the concept of precipitating factors in the onset of myodegeneration in this disorder. The nature of these precipitating factors has been the subject of some discussion. Allen et al. [32] have suggested that a sudden increase in muscular activity and inclement weather were precipitating factors such that the animals became selenium- or vitamin E-deficient whilst housed indoors on an inadequate diet during the winter, but the disease did not become apparent until the animals were stressed on 'turnout'. Anderson et al. [33] further studied the effect of exercise on the onset of myodegeneration in calves at turnout

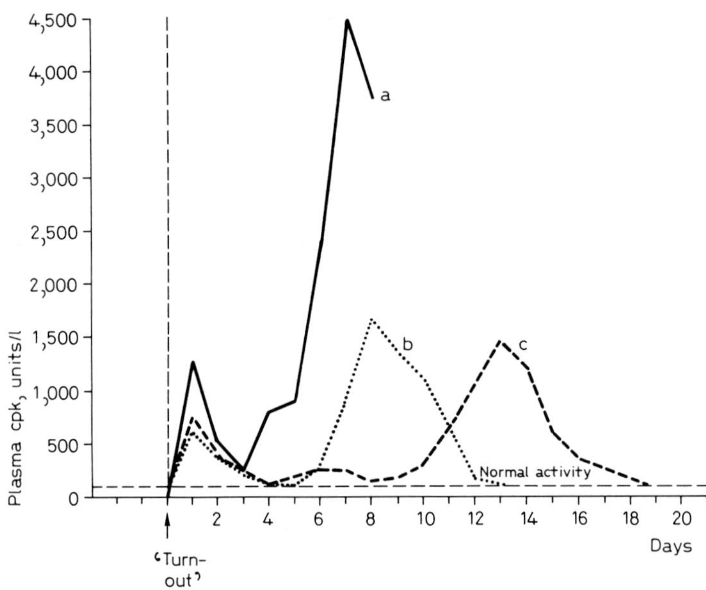

Fig. 1. Changes in plasma creatine kinase activity (plasma CPK) in 3 vitamin E-depleted calves following 'turnout' to pasture. The normal activity in cattle is about 50 IU/l. Reproduced with permission from Anderson et al. [33].

following a period of feeding indoors with a vitamin E-deficient diet. They monitored the appearance of the muscle specific enzyme creatine kinase in the blood of these animals following turnout and found those effected by the myodegenerative process had a biphasic rise in their plasma creatine kinase activity (fig. 1). Almost all of the animals, whether or not they were vitamine E-deficient, showed the first small rise in plasma creatine kinase activity which peaked at 1–2 days post turnout, but only those which eventually developed severe myodegeneration had a large second rise in plasma creatine kinase activity.

The implication of this work was that an initial stress (in this case exercise) initiated a small amount of muscle degeneration which the vitamin E-replete calves could withstand and were able to repair, whereas the vitamin E-deficient animals were unable to prevent the exascerbation of this damage to produce significant myodegeneration.

An alternative precipitating agent was suggested by McMurray et al. [34], who found that feeding of polyunsaturated fatty acids such as linoleic (18:2) or linolenic (18:3) acids to calves maintained indoors on selenium- or vitamin E-deficient diets induced a rapid onset of myopathy and suggested that turnout also allows the calves access to newly grown grass which contains large amounts of linolenic and linoleic acids. McMurray et al. [34] suggested that this leads to an increased amount of substrate for lipid peroxidation within the muscles. The lack of selenium or vitamin E then predisposes the animals to enhanced membrane degradation by lipid peroxidation.

It is likely that all of these factors and others can predispose the muscle of selenium- or vitamin E-deficient animals to myopathy. Jackson et al. [21] have demonstrated that isolated muscle from vitamin E-deficient animals are more susceptible to contractile activity induced damage than vitamin E-replete muscles while Kakulus [35] has examined the factors affecting the onset of myopathy in the vitamin E-deficient marsupial, the quokka, and has found this to be influenced by the size of the enclosure in which the animal is held.

Several workers have examined muscle of vitamin E- and selenium-deficient animals for direct evidence of the involvement of free radicals in the pathogenesis of the myopathy. Brady et al. [17] found elevated levels of malonaldehyde in muscle of animals with vitamin E and selenium deficiency, while Arthur and Morrice [20] and Jackson et al. [21] found increased amounts of malonaldehyde were produced by vitamin E-deficient muscle during incubation 'in vitro', suggesting a decrease in the ability of these muscles to inhibit lipid peroxidation. Arthur and Morrice [20] also studied the efflux of ethane from muscles deficient in either vitamin E or vitamin E and selenium 'in vitro' and found this to be elevated although the same was not true of muscles deficient in selenium alone. Finally in this area Davies et al. [12] have examined the stable free radical signal obtained by ESR spectrometry from vitamin E-deficient rat muscle 'in vitro' and found it to be enhanced compared to vitamin E-sufficient rat muscle.

An alternative approach to examining the role of free radicals in vitamin E and selenium deficiency myopathy was also undertaken by Arthur and Morrice [20] who reasoned that if free radicals were involved in the process of damage then substances which enhance free radical production should induce an increase in the amount of myopathy in these animals. It had previously been shown that intraperitoneal injections of iron salts could induce myopathic changes in vitamin E-deficient rat muscle [25] and Arthur and Morrice used this as a stress in animals of known vitamin E and selenium status. They examined the plasma activity of the muscle-derived enzymes creatine kinase and pyruvate kinase as an index of the rate of progress of myopathy and found that intraperitoneal iron injections induced a large rise in the plasma pyruvate kinase activity and a smaller rise in plasma creatine kinase activity in the animals deficient in vitamin E or vitamin E and selenium, but had little effect on the animals deficient in selenium alone.

In summary it therefore appears that there is some evidence that free radicals are involved in the pathogenesis of vitamin E and selenium deficiency myopathy in animals. However the data in support of this are much stronger in the case of vitamin E deficiency than selenium deficiency and further work is required in this area.

The situation with human vitamin E and selenium deficiency is much less clear. The only human disease known to be responsive to vitamin E and probably caused by vitamin E deficiency are the neurological changes associated with abeta-lipoproteinaemia [36]. Vitamin E is carried in the blood associated with beta-lipoproteins and hence there is a defect in vitamin E transport in this disease. There is little evidence of muscular involvement in this disease, although the plasma creatine kinase activity is mildly elevated in many patients [36]. Selenium deficiency has been described in patients during long-term total parenteral nutrition and may be associated with myopathy [37, 38]. In addition a cardiomyopathy in China (Keshan disease) has been ascribed to dietary selenium deficiency [39] although there does not appear to have been any work designed to examine whether free radicals are involved in the pathogenesis of this disease.

Free Radicals in the Pathogenesis of the Human Muscular Dystrophies

The previously described similarities between vitamin E and selenium deficiency myopathy in animals and the human muscular dystrophies has prompted many workers to investigate the possibility that selenium or vitamin E deficiency could play a role in these human diseases, and in addition the recognition that these agents are important in the protective mechanisms that the body has developed against damaging free radicals species has suggested studies to examine the possibility that free radical mechanisms play a role in the pathogenesis of the muscular dystrophies.

The situation concerning the possible role of abnormalities in either selenium and vitamin E metabolism in the most severe (Duchenne) form of muscular dystrophy is somewhat confused. Westermark [40] has claimed that plasma selenium levels are abnormally low in patients with Duchenne muscular dystrophy, and has also found an abnormally rapid turnover of selenium in patients with this disease [41]. However, we have been unable to demonstrate any abnormality in selenium metabolism in these diseases [42]. Other workers have suggested that oral selenium therapy may be beneficial in myotonic muscular dystrophy [43, 44]. Formal double-blind trials of oral selenium supplements in patients with muscular dystrophy are now being undertaken in the UK which should resolve this confusion [45].

Table 1. Malonaldehyde or thiobarbituric-reacting substances in muscle from patients with muscular dystrophy

	Controls	Duchenne muscular dystrophy patients
Kar and Pearson [15],	0.70 ± 0.10	1.67 ± 0.20
A530/g wet wt	(6)	(8)
Jackson et al. [16],	393.0 ± 58.9	1340.6 ± 312.4
n moles malonaldehyde/g protein	(11)	(7)
Mechler et al. [13],	1.33 ± 0.16	3.87 ± 0.71
A530/mg protein	(5)	(4)

The numbers of patients and control samples are given in parentheses.

We have also found normal plasma tocopherol levels in patients with Duchenne muscular dystrophy [16] which is in contrast to Hunter et al. [8] and Austin et al. [46] who found plasma tocopherol levels were low in these patients. In addition it has been reported that plasma lipoproteins may be abnormal in patients with Duchenne muscular dystrophy [46] and therefore that the plasma carrier for vitamin E may be abnormal, i.e. a situation analagous to that which occurs in abeta-lipoproteinaemia. Early workers reported that vitamin E therapy was of benefit to patients with Duchenne muscular dystrophy [47] but carefully controlled trials have demonstrated that there is no beneficial effect of short-term vitamin E treatment in these diseases [48, 49].

Several groups of workers have examined body fluids of patients with muscular dystrophy for evidence of abnormal production of free radical products or changes in free radical protective mechanisms [8–11] and have concluded that there may be some abnormality in this disease. In addition measurements of malonaldehyde (or thiobarbituric acid reactive materials) in muscle diopsy samples, from patients with Duchenne muscular dystrophy, have suggested that increased lipid peroxidation may occur in muscle of patients with this disorder (table 1).

Supportive data has also been obtained from studies of animal models, such as avian muscular dystrophy [50], but it should be noted that none of the animal models of this disease are exact models of the human disease and should ideally only be thought of as animal models of genetically inherited, chronic muscle wasting disorders. The fact that similar findings of increased malonaldehyde are found in these species suggest that the changes may not be primary in the pathogenesis of the human muscular dystrophies, but are secondary to muscle damage.

It must also be noted that analysis of muscle from either patients with muscular dystrophy or animal models of the disease is itself subject to severe problems. The muscle samples obtained are likely to contain a considerable accumulation of fat and collagen which infiltrate the muscles in these diseases. Although it is possible to make allowance for reduced muscle mass by simultaneous measurements of muscle-specific standards such as non-collagen protein or actin, this does not remove the possibility of interference in the analysis in question by the increased intracellular fat or collagen content.

In conclusion it appears that there is no major abnormality of vitamin E or selenium metabolism in patients with muscular dystrophy, although there may be minor changes in both. There is evidence of increased free radical activity in dystrophic muscle both human and animal, but the relevance of this is to the pathogenesis of the disease is unclear, since it may be a consequence rather than a cause of muscle degeneration.

Free Radicals and Exercise

The possibility that free radicals might be generated in the muscle during exercise was first suggested by Dillard et al. [1] who examined the effect of exercise and vitamin E on pentane excretion in man. In this work they examined human subjects and found that there was a significant increase in breath pentane excretion during exercise at 50% of the subject's maximum oxygen uptake and that this was partially inhibited by oral vitamin E supplements. The experimental procedure which was used was extremely complex since background levels of pentane in the air were very high. The subjects therefore had to breathe hydrocarbon-scrubbed air prior to undertaking the exercise in order to 'wash out' the lungs of the background pentane and the whole experiment was undertaken in a closed system so that room air was not taken in by the subject. This experiment was later repeated in animals by Gee and Tappel [4] who confirmed the finding of Dillard et al. [1]. These experiments stimulated a number of groups of workers to further study the role of free radicals in exercise-induced damage to tissue, but because of the difficulties in the experimental techniques the studies of Dillard and co-workers remained unrepeated until 1986 when Snider et al. [5] reported similar findings. However they also reported that the amount of the environmental contaminant isopentane in exhaled air during exercise was also significantly elevated. The reasons for this and the implications have previously been discussed, but in essence it appears that measurements of exhaled hydrocarbons during exercise are an unsuitable way of monitoring lipid peroxidation.

Other evidence for an increase in lipid peroxidation during exercise has been presented by Brady et al. [17], who examined the malonaldehyde content of tissues following swimming exercise in the rat. They found that this stress increased the malonaldehyde content of muscle and other tissues. Davies et al. [12] also found that the malonaldehyde content of muscle was increased following running exercise in the rat and additionally studied the amplitude of the stable free radical signal seen in normal tissue by ESR techniques. They found that this also increased in both muscle and liver following exercise.

As supporting evidence for the thesis that free radical activity is increased in tissues during exercise, Packer and co-workers [51, 52] have also examined the effect of antioxidant (vitamin E and vitamin C) deficiency on exercise endurance in experimental animals. They found that vitamin E deficiency leads to a severe reduction in exercise endurance in the rat. This is associated with a loss of mitochondrial oxidative capacity which these workers attributed to increased oxidative damage to the muscle mitochondria [51]. They have also demonstrated gross changes in cellular glutathione metabolism during exercise which may have been induced by increased free radical activity and may be responsible for some of the pathological damage which occurs during exercise.

In summary it is therefore unclear whether increased free radical activity occurs during exercise. The finding that the increased pentane excretion seen during exercise may be artefactual has undermined what had begun to appear to be a relatively clear picture. The reported increase in malonaldehyde content of muscle and liver during exercise suffers from the problems previously explained in terms of the increased levels in muscle biopsies from muscular dystrophy patients while the gross changes in exercise endurance formed in antioxidant-deficient animals provides only supportive evidence since this presupposes that the most important role of vitamin E or vitamin C in tissues during exercise is as scavengers of free radicals.

Free Radicals and Muscle Damage

Damage to skeletal muscle occurs in a number of different physiological and pathological situations such as following excessive or unaccustomed exercise and in patients with degenerative muscle disorders such as the muscular dystrophies or polymyositis. In an attempt to find rational ways to reduce muscle damage and help maintain cell viability in disorders such as muscular dystrophy, we have been studying the mechanisms by which loss of cell viability occurs following a variety of different stresses. Surprisingly this work has suggested that free radical-mediated processes may be involved in a general mechanism by which loss of cell viability occurs following a number of different damaging stimuli. This may suggest a role which free radicals play in skeletal muscle pathology of various different types and may explain some of the apparently discrepant findings of the role of free radicals in the various forms of skeletal muscle pathology which have been previously described.

Our studies of damage to skeletal muscle have primarily utilised an 'in vitro' model system in which mouse or rat extensor digitorum longus or soleus muscles are incubated in a bicarbonate-buffered medium [53]. The apparatus is designed such that the muscles can be damaged by excessive contractile activity or by addition of toxins or poisons to the medium. The efflux of intracellular enzymes is routinely used as an index of the damage to the muscles, but the muscles can also be taken for examination of their ultrastructural appearance using electron microscopy [54].

These studies initially indicated that two different types of biochemical processes might be involved in damage to skeletal muscle, namely calcium-activated degenerative mechanisms and free radical-mediated oxidation of membrane lipids, but our recent work suggests that these are likely to be part of the same process.

Evidence for a Role of Calcium in Skeletal Muscle Damage. Changes in the intracellular calcium content have been implicated in the mechanisms by which damage occurs to several tissues including skeletal muscle. In cardiac tissue damage due to hypoxia or re-oxygenation has been shown to be associated with an increase in tissue calcium content [55], while in hepatocytes loss of cell viability following incubation with various toxins had been reported to be dramatically reduced when the external calcium is removed from the incubation fluid [56], this finding has subsequently been the subject of much controversy [57, 58].

High external calcium concentrations (3–10 mmol/l) have been shown to increase creatine kinase release from resting animal [59] and human skeletal muscle [60]. Treatment of skeletal muscle preparations with the calcium ionophore, A23187, has further demonstrated the potential of increased calcium levels to induce damage [61]. Calcium channel-blocking agents (i.e. calcium antagonists) have been found to reduce the creatine kinase released from human skeletal muscle 'in vitro' [62], and other workers have shown that alternative manipulations designed to reduce calcium accumulation in skeletal muscle (i.e. parathyroidectomy) prevents the pathological changes to skeletal muscle in hampsters with congenital forms of muscular dystrophy [63].

In experiments with the 'in vitro' muscle damage model it was found that release of enzymes following different stresses (e.g. excessive contractile activity, treatment with low-dose detergents or treatment with mitochondrial inhibitors) could be prevented by removal of the external calcium during the damaging procedure [54, 64]. It was also found that this manipulation was equally effective in protection of muscle against the histological and electron microscopic changes induced by excess contractile activity [54]. Other experiments with this system have also demonstrated a dramatic increase in total muscle calcium during either excess contractile activity leading to enzyme efflux or treatment with mitochondrial inhibitors (e.g. dinitrophenol [65]).

These results suggest that damage to skeletal muscle is accompanied by an influx of extracellular calcium down the large extracellular to intracellular concentration gradient for this element. This increased intracellular calcium content then mediates further pathological changes. Despite the fact that others have claimed an effect of calcium antagonists on skeletal muscle [62], we have been unable to demonstrate any protective effect of these agents in isolated preparations [54].

Considerable speculation has surrounded the possible mechanisms by which an increased intracellular calcium content mediates pathological processes in cells. The hypotheses which have been proposed regarding skeletal muscle include accumulation of calcium by mitochondria leading to loss of oxidative energy production [66], activation of calcium-dependent proteases [67], activation of proteases via the stimulation of prostaglandin production [68] or direct release of lysosomal enzymes [69].

Inhibitor studies which we have performed suggest that calcium activation of phospholipase A may be a key step in the damage leading to enzyme efflux [64]. Activation of this enzyme will result in the breakdown of membrane phospholipids leading to production of lysophospholipids and free fatty acids. Accumulation of lysophospholipids could lead to a breakdown of membrane lipid organisation [70] and the free fatty acids released could act as detergents causing membrane damage [71]. In addition among the free fatty acids produced will be arachidonic acid. This is the precursor for the prostaglandin series of compounds. Prostaglandins have been reported to be involved in the control of muscle protein turnover [68, 72, 73]. In recent experiments we have demonstrated that there is indeed a large release of prostaglandins E_2 and $F_{2\alpha}$ in association with the efflux of intracellular enzymes [74]. These studies therefore support the hypothesis that calcium activates phospholipase enzymes to cause release of arachidonic acid which can then be converted to prostaglandins via the cyclo-oxygenase pathway.

Evidence for a Role of Free Radicals in Skeletal Muscle Damage. Several workers have proposed that free radicals are produced in skeletal muscle during exercise [1, 12, 17, 51], but evidence that free radicals are involved in the process of damage following this and other stresses is

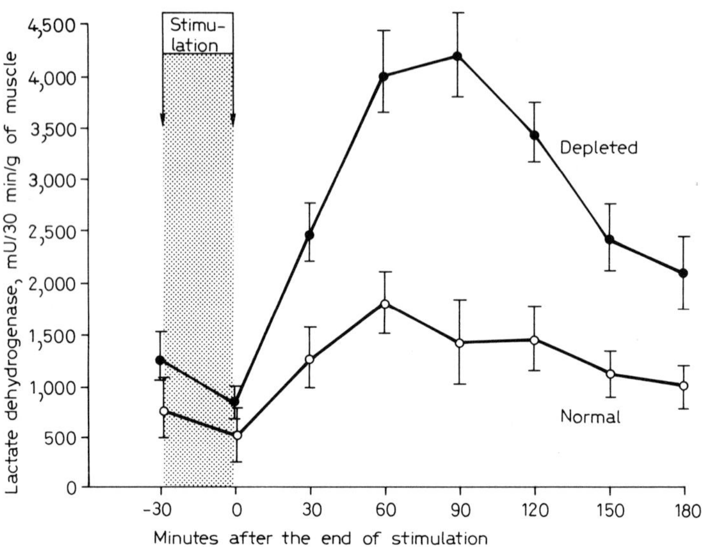

Fig. 2. Efflux of intracellular lactate dehydrogenase from soleus muscles of mice fed a low vitamin E diet (●) or the same diet supplemented with vitamin E(○) in response to electrical stimulation at 100 Hz for 0.5 s every 5 s for 30 min in oxygenated medium. The efflux from the muscles of the vitamin E-depleted mice was significantly ($p < 0.001$) greater than that from the control group, at all times after the end of stimulation. Reprinted with permission from Jackson et al. [21].

sparse. We have demonstrated a transient rise in the malonaldehyde content of skeletal muscle following damaging contractile activity [75] and have also demonstrated that the vitamin E content of the muscle influences the amount of intracellular enzymes released from the muscles following an equivalent amount of contractile activity [21] both 'in vitro' and 'in vivo' (fig. 2). ESR studies have also suggested that an increase in the free radical activity of muscle is associated with damage to the muscle. We have described a $70 \pm 20\%$ increase in the stable ESR signal in muscle in which leakage of intracellular enzymes was induced by excessive contractile activity 'in vivo' [24] (fig. 3).

Link between Calcium- and Free Radical-Induced Damage. We have previously proposed that calcium accumulation may produce increased free radical damage to muscles since the free fatty acids released from membranes is more susceptible to free radical-mediated lipid peroxidation [76]. The rationale behind this was that while an integral part of the membrane

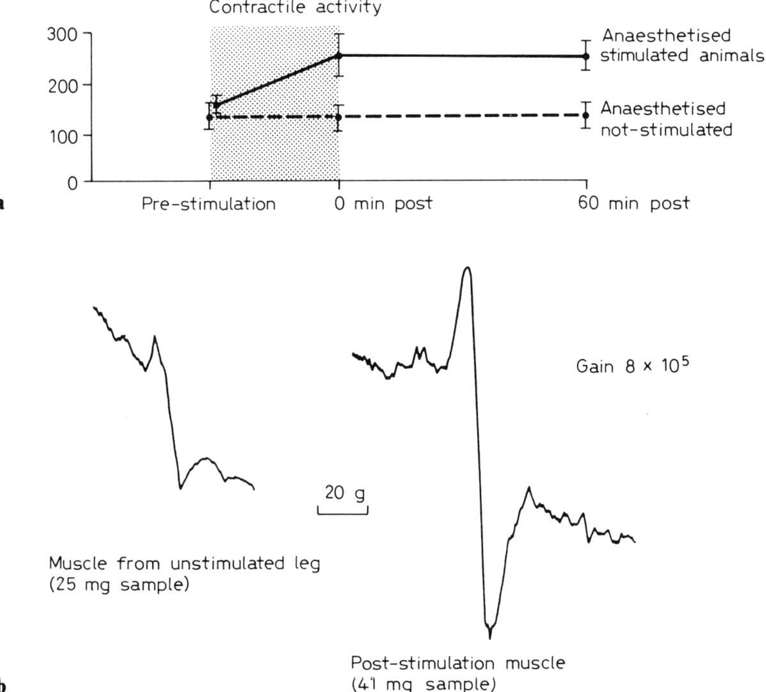

Fig. 3. a Plasma creatine kinase activity of a group of anaesthetised male Wistar rats in whom the muscles of one hind limb were electrially stimulated (at 50 Hz and 70V for 0.5 s repeated every 2 s) for 30 min compared to a group of animals which were anaesthetised but not stimulated. **b** ESR signal (g = 2.004) from gastrocnemius muscles with and without 30 min electrical stimulation inducing excessive contractive activity. Muscles were removed immediately following the end of stimulation for ESR examination. Reprinted with permission from Jackson et al. [24].

phospholipids the polyunsaturated fatty acids were associated with, and protected by, vitamin E, whereas once released there will be no similar cytoplasmic lipid-soluble antioxidants available to protect the molecule. In addition it was felt that the cytoplasm was the major site of free radical production since several enzyme systems produce reactive oxygen metabolites as a normal component of metabolism. Non-enzymatic peroxidation of the liberated fatty acid would therefore result.

Our recent studies suggest that the link between a role of calcium and free radicals may be an enzymatic system. Inhibitors of lipoxygenase enzymes have been found to have a potent protective effect against damage

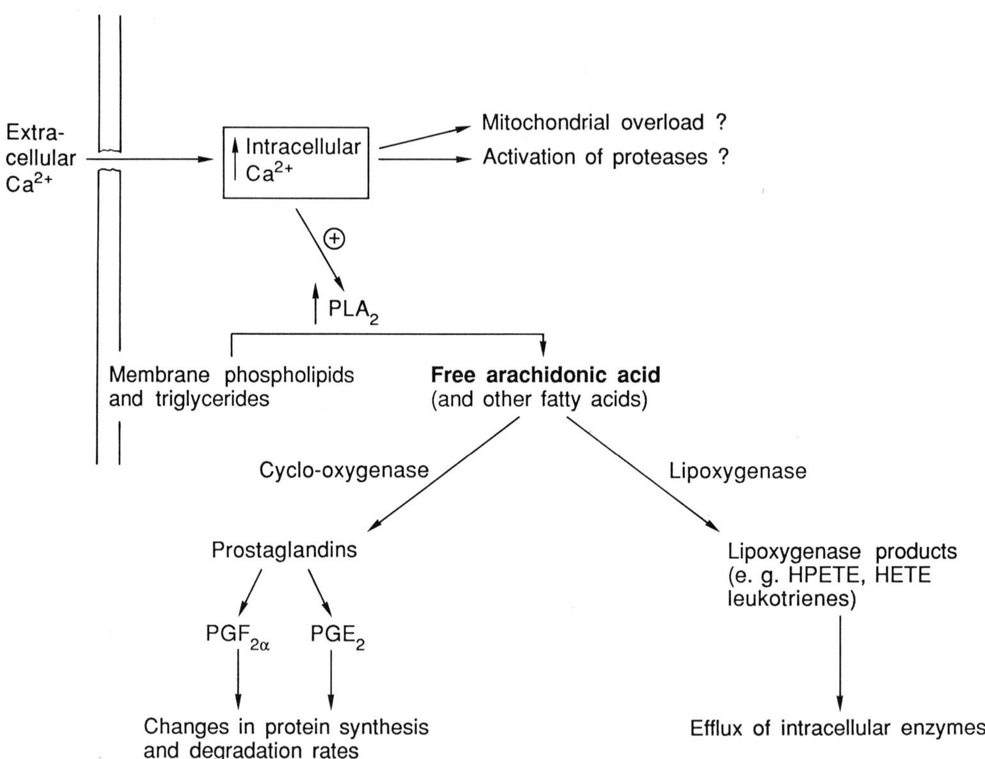

Fig. 4. Proposed mechanisms of skeletal muscle damage. It is suggested that a number of different stresses induce loss of cell viability in skeletal muscle by a common mechanism involving an elevation of intracellular calcium levels. This leads to an activation of phospholipase A2 (PLA2) enzymes and a consequent release of arachidonic acid and other polyunsaturated lipids from membrane lipids. The arachidonic acid is then converted to prostaglandins via cyco-oxygenase leading to changes in protein turnover rates [68, 72, 73] and peroxidised, either non-enzymatically or enzymatically via lipoxygenase to products which induce sarcolemmal changes and an efflux of intracellular components.

to muscle induced by the calcium ionophore, A23187 [74], suggesting that excess calcium induces release of arachidonic acid via phospholipase enzymes and this released acid is then converted to other metabolites via lipoxygenase. It therefore follows that some product of the lipoxygenase pathway may induce the changes in the muscle sarcolemma which lead to intracellular enzyme efflux.

However, it would also be noted that the evidence for lipoxygenase involvement in skeletal muscle damage is only preliminary and entirely dependent upon the action of inhibitors. Since it is unlikely that any such inhibitors are specific to only one enzyme products of lipoxygenase activity must be demonstrated in skeletal muscle before a role can be attributed to this enzyme. Lipoxygenase enzymes have not been described in skeletal muscle tissue, but are present in smooth [77] and possibly cardiac [78] muscle. A schematic diagram of the proposed mechanisms of skeletal muscle damage is shown in figure 4.

Overview of the Role of Free Radicals in Skeletal Muscle Pathology

Almost all of the findings regarding free radicals in skeletal muscle pathology are explainable in terms of the mechanisms of damage hypothesised in figure 4. Thus in vitamin E and selenium deficiency myopathy there is evidence that a small amount of damage precedes the onset of severe myopathy. Figure 1 shows the plasma creatine kinase activity of calves at turnout [33] and it is apparent that there is an initial small amount of damage to skeletal muscle which occurs in all animals, in response to unaccustomed exercise, but then those deficient in vitamin E go on to develop myopathy whereas the vitamin E-replete animals do not. It could be envisaged that vitamin E-sufficient animals have the ability to prevent further damage from the lipid peroxidation products produced by the mechanism described in figure 4 whereas vitamin E-depleted animals do not. Alternatively it has been reported that vitamin E inhibits lipoxygenase enzymes [79, 80] at normal tissue concentrations and it may be that the initial problems in depleted animals arise from an overactivity of muscle lipoxygenase during vitamin E deficiency.

In selenium deficiency accumulation of calcium has been reported to proceed the biochemical, histological or clinical evidence of myopathy [81] and it may be that this leads to an increase in lipid peroxides via the mechanisms proposed which the cell is unable to detoxify because of a lack of the selenium-dependent enzyme, glutathione peroxidase.

It has been proposed that Duchenne muscular dystrophy results from a defect in the muscle cytoplasmic membrane [82, 83] and that one of the main consequences of this is an ingress of extracellular calcium [84–86]. Muscle calcium levels have been reported to be elevated in muscular dystrophy by several workers [84–87] and muscle phospholipase A activity is reported to be grossly abnormal [88]. This suggests that the scheme outlined in figure 4 is activated in dystrophic muscle and would explain the elevated content of malonaldehyde found in this tissue (table 1) and the increase in other indices of lipid peroxidation found in plasma of patients. It is therefore not necessary to propose increased free radical activity as primary to the disease since it could result as a consequence of the well-known calcium accumulation in dystrophic muscle.

The situation in exercise is similar, it is important to separate the effects of exercise per se from the effects of exercise-induced damage, since once damage is initiated production of increased amounts of lipid peroxides and increased free radical activity will result. Most of the studies in which elevated levels of peroxidation products have been demonstrated in muscle following exercise have involved experiments in which animals were exercised to exhaustion [17] and in these circumstances exercise-induced damage may have resulted.

In summary it is apparent that most of the reported association between free radicals and pathology in skeletal muscle can be explained if it is accepted that lipid peroxidation (possibly via an enzymatic mechanism) plays a role in a general mechanism of muscle damage common to a number of different situations. It is important to stress that this is an active role in the process of damage and that increased lipid peroxidation occurs prior to the loss of cell viability. Substances which might inhibit (enzymatic) lipid peroxidation might therefore reduce muscle damage in all of the situations cited here.

Acknowledgements

The author would like to acknowledge the excellent technical assistance of Miss C. Forte, Miss L. Burns, Miss J. Roberts and Mrs. S. Page and helpful discussions and collaboration from Prof. R.H.T. Edwards, Dr. D.A. Jones, Prof. M. Symons, Prof. A.T. Diplock and Dr. A.J.M. Wagenmakers. Continuing financial support from the Muscular Dystrophy Group of Great Britain and F. Hoffman-La Roche & Co. is also gratefully acknowledged.

References

1 Dillard CJ, Litov RE, Savin WM, et al: Effects of exercise, vitamin E and ozone on pulmonary function and lipid peroxidation. J Appl Physiol 1978;45:927–932.
2 Tappel AL, Dillard CJ: 'In vivo' lipid peroxidation measurement via exhaled pentane and protection by vitamin E. Fed Proc 1981;40:174–178.
3 Hafeman DG, Hoekstra WG: Protection against carbon tetrachloride-induced lipid peroxidation in the rat by dietary vitamin E, selenium and methionine as measured by ethane evolution. J Nutr 1977;107:656–665.
4 Gee DL, Tappel AL: The effect of exhaustive exercise on expired pentane as a measure of 'in vivo' lipid peroxidation in the rat. Life Sci 1981;28:2425–2429.
5 Snider MT, Balke PO, Oertes KE, et al: Lipid peroxidation during muscular exercise in

man: inferences from the pulmonary excretion of n-pentane, isopentane and nitrogen (abstract). Proc Nutr Soc 1986;45:64A.

6   Recknagel RO, Ghoshal AK: Quantitative estimation of peroxidative degeneration of rat liver microsomal and mitochondrial lipids after carbon tetrachloride poisoning. Exp Mol Pathol 1966;5:413–426.

7   Fletcher BL, Dillard CJ, Tappel AL: Measurement of fluorescent lipid peroxidation products in biological systems and tissues. Anal Biochem 1973;52:1–9.

8   Hunter MIS, Mohamed JB: Plasma antioxidants and lipid peroxidation products in Duchenne muscular dystrophy. Clin Chim Acta 1986;155:123–132.

9   Hunter MIS, Brzeski MS, de Vane PJ: Superoxide dismutase, glutathione peroxidase and thiobarbituric acid-reactive compounds in erythrocytes in Duchenne muscular dystrophy. Clin Chim Acta 1981;115:93–98.

10   Matkovics B, Laszlo A, Szabo L: A comparative study of superoxide dismutase, catalase and lipid peroxidation in red blood cells from muscular dystrophy patients and normal controls. Clin Chim Acta 1982;118:289–292.

11   Burri BJ, Chan SG, Berry AJ, et al: Blood levels of superoxide dismutase and glutathione peroxidase in Duchenne muscular dystrophy. Clin Chim Acta 1980;105:249–255.

12   Davies KJA, Quintanilha AT, Brooks GA, et al: Free radicals and tissue damage produced by exercise. Biochem Biophys Res Commun 1982;107:1198–1205.

13   Mechler F, Imre S, Dioszeghy P: Lipid peroxidation and superoxide dismutase in Duchenne muscular dystrophy. J Neurol Sci 1984;63:279–283.

14   Edwards RHT, Round JM, Jones DA: Needle biopsy of skeletal muscle: A review of 10 years experience. Muscle Nerve 1983;6:676–683.

15   Kar NC, Pearson CM: Catalase, superoxide dismutase, glutathione reductase and thiobarbituric acid-reactive products in normal and dystrophic human muscle. Clin Chim Acta 1979;94:277–280.

16   Jackson MJ, Jones DA, Edwards RHT: Techniques for studying free radical damage in muscular dystrophy. Med Biol 1984;62:135–138.

17   Brady PS, Brady LJ, Ullrey DE: Selenium, vitamin E and the response to swimming stress in the rat. J Nutr 1979;109:1103–1109.

18   Halliwell B, Gutteridge JMC: Lipid peroxidation, oxygen radicals, cell damage and antioxidant therapy. Lancet 1984;ii:1396–1397.

19   Siu GM, Draper HH: Metabolism of malonaldehyde 'in vivo' and 'in vitro'. Lipids 1982;17:349–355.

20   Arthur JR, Morrice PC: The effects of Se and vitamin E deficiencies on the response of rats to Fe injection; in Mills CF, Bremner I, Chesters JK (eds): Proc TEMA V, Aberdeen. Slough, Commonwealth Agriculture Bureau, 1986.

21   Jackson MJ, Jones DA, Edwards RHT: Vitamin E and skeletal muscle; in Porter R, Whelan J (eds): Ciba Fdn Symp Ser No 101, Biology of vitamin E. London, Pitman, 1983;pp.224–239.

22   Commoner B, Townsend J, Pake GE: Free radicals in biological materials. Nature 1954;174:689–691.

23   Heckley RJ: Free radicals in dry biological systems; in Pryor WA (ed): Free Radicals in Biology. New York, Academic Press, 1976;vol 2,pp.135–158.

24   Jackson MJ, Edwards RHT, Symons MCR: Electron spin resonance studies of intact mammalian skeletal muscle. Biochem Biophys Acta 1985;847:185–190.

25   Dougherty JJ, Croft WA, Hoekstra WG: Effects of ferrous chloride and iron-dextran on lipid peroxidation 'in vivo' in vitamin E and selenium adequate and deficient rats. J Nutr 1981;111:1784-1796.

26   Okabi E, Hiyama E, Oyama M, et al: Free radical damage to sarcoplasmic reticulum of masseter muscle by arachidonic acid and prostaglandin G_2. Pharmacology 1982;25:138–148.

27   Harris EJ, Booth R, Cooper MB: The effect of superoxide generation on the ability of mitochondria to take up and retain Ca^{2+}. FEBS Lett 1982;146:267–272.

28   Jackson MJ, Jones DA: Unpublished observations.

29   Wolff SP, Garner A, Dean RT: Free radicals, lipids and protein degradation. TIBS 1986;Jan:27–31.

30   Bradley R, Fell BF: Myopathies in animals; in Walton JN (ed): Disorders of Voluntary Muscle, ed 4. London, Churchill-Livingstone, 1980;pp.824–872.

31   Allen WM, Parr WH, Bradley R: Loss of vitamin E in stored cereals in relation to a myopathy of yearling cattle. Vet Rec 1974;94:373–375.

32   Allen WM, Bradley R, Berrett S, et al: Degenerative myopathy with myoglobinuria in yearling cattle. Br Vet J 1975;131:292–308.

33   Anderson PH, Bradley R, Barrett S, et al: The sequence of myodegeneration in nutritional myopathy of the older calf. Br Vet J 1977;133:160–165.

34   McMurray CH, Rice DA, Kennedy S: Experimental models for mutritional myopathy; in Porter R, Whelan J (eds): Ciba Fdn Ser No 101, Biology of Vitamin E. London, Pitman, 1983;pp.201–223.

35   Kakulus BA: Influence of the size of enclosure on the development of myopathy in the Rottnest Quokka. Nature 1963;198:673–674.

36   Muller DPR, Lloyd JK, Wolff OH: Vitamin E and neurological function: abetalipoproteinaemia and other disorders of fat absorption; in Porter R, Whelan J (eds): Ciba Fdn Symp Ser No 101, Biology of Vitamin E. London, Pitman, 1983;pp.106–121.

37   Van Rij AM, Thomson CD, McKenzie JM, et al: Selenium deficiency in total parenteral nutrition. Am J Clin Nutr 1979;32:2076–2085.

38   Brown MR, Cohen HJ, Lyons JM, et al: Proximal muscle weakness and selenium deficiency associated with long term parenteral nutrition. Am J Clin Nutr 1986;43:549–554.

39   Keshan Disease Research Group of the Chinese Academy of Medical Sciences: Observations on the effect of sodium selenite in prevention of Keshan disease. Chin Med J 1979;93:471–476.

40   Westermark T: Selenium content of tissues in Finnish infants and adults with various diseases, and studies on the effects of selenium supplementation in neuronal ceroid lipofuscinosis patients. Acta Pharmacol Toxicol 1977;41:121–128.

41   Westermark T, Rahola T, Kallio A-K, et al: Long-term turnover of selenite-Se in children with motor disorders. Klin Paediatr 1982;194:301–302.

42   Jackson MJ, Round JM, Diplock AT, et al: Selenium status of patients with Duchenne muscular dystrophy (abstract). Eur J Clin Invest 1985;15:138.

43   Orndahl G, Rindby A, Selin E: Myotonic dystrophy and selenium. Acta Med Scand 1984;211:493–499.

44   Orndahl G, Sellden U, Hallin S, et al: Myotonic dystrophy treated with selenium and vitamin E. Acta Med Scand 1986:219:407–414.

45   Edwards RHT, Coakley J, Stokes M, et al: A tissue pharmacology approach to therapeutic trials in muscular dystrophy (abstract). Muscle Nerve 1986;9(suppl):270.

46   Austin L, Arthur H, Ionello R, et al: Lipoproteins in plasma of Duchenne muscular dystrophy patients and female carriers; in Kidman AD (ed): Molecular Pathology of Nerve and Muscle. 3rd Symp Fdn for Life Sciences, Sydney, Australia. New York, Humana Press, 1984;pp.385–395.

47 Bicknell F: Vitamin E in the treatment of muscular dystrophies and nervous diseases. Lancet 1940;i:10–13.

48 Fitzgerald C, McArdle B: Vitamins E and B_6 in the treatment of muscular dystrophy and motor neurone disease. Brain 1941;64:19–42.

49 Walton JN, Nattrass FJ: On the classification, natural history and treatment of the myopathies. Brain 1954;77:169–231.

50 Omaye ST, Tappel AL: Glutathione peroxidase, glutathione reductase, and thiobarbituric acid-reactive products in muscles of chickens and mice with genetic muscular dystrophies. Life Sci 1975;15:137–145.

51 Quintanilha AT, Packer L: Vitamin E, physical exercise and tissue oxidative damage; in Porter R, Whelan J (eds): Ciba Fdn Symp Ser No 101: Biology of Vitamin E. London, Pitman, 1983;pp.56–69.

52 Packer L: Vitamin E, physical exercise and tissue damage in animals. Med Biol 1984;62:105–109.

53 Jones DA, Jackson MJ, Edwards RHT: Release of intracellular enzymes from an isolated mammalian skeletal muscle preparation. Clin Sci 1983;65:193–201.

54 Jones DA, Jackson MJ, McPhail G, et al: Experimental muscle damage: The importance of external calcium. Clin Sci 1984;66:317–322.

55 Nayler WG, Poole-Wilson PA, Williams A: Hypoxia and calcium. J Mol Cell Cardiol 1979;11:683–706.

56 Schanne FX, Kane AB, Young AB, et al: Calcium dependence of toxic cell death: a final common pathway. Science 1979;206:700–701.

57 Smith MT, Thor H, Orrenius S: Toxic injury to isolated hepatocytes is not dependent on extracellular calcium. Science 1981;213:1257–1259.

58 Farris MW, Pascoe GA, Reed DJ: Vitamin E reversal of the effect of extracellular calcium on chemically induced toxicity in hepatocytes. Science 1985;227:751–754.

59 Soybell D, Morgan J, Cohen L: Calcium augmentation of enzyme leakage from mouse skeletal muscle and its possible site of action. Res Commun Chem Pathol Pharmacol 1978;20:317–329.

60 Anand R, Emery AEH: Calcium stimulated enzyme efflux from human skeletal muscle. Res Commun Chem Pathol Pharmacol 1980;28:541–550.

61 Duncan CJ: Role of intracellular calcium in promoting muscle damage: a strategy for controlling the dystrophic condition. Experientia 1978;34:1531–1535.

62 Anand R, Emery AEH: Verapamil and calcium-stimulated enzyme efflux from skeletal muscle. Clin Chem 1982;28:1482–1484.

63 Palmieri GMA, Nutting DF, Bhattacharya SK, et al: Parathyroid ablation in dystrophic hamsters. J Clin Invest 1980;68:646–654.

64 Jackson MJ, Jones DA, Edwards RHT: Experimental skeletal muscle damage: the nature of the calcium-activated degenerative processes. Eur J Clin Invest 1984;14:369–374.

65 Claremont D, Jackson MJ, Jones DA: Accumulation of calcium in experimentally damaged mouse muscles (abstract). J Physiol 1984;353:57P.

66 Wrogemann K, Pena SJG: Mitochondrial overload: a general mechanism for cell necrosis in muscle diseases. Lancet 1976;ii:672–674.

67 Ebashi S, Sugita H: The role of calcium in physiological and pathological processes of skeletal muscle; in Aguayo AJ, Karpati G (eds): Current Topics in Nerve and Muscle Research. Amsterdam, Excerpta Medica, 1979;pp.73–84.

68 Rodemann HP, Waxman L, Goldberg AL: The stimulation of protein degradation in muscle by Ca^{2+} is mediated by prostaglandin E_2 and does not require the calcium-activated protease. J Biol Chem 1981;257:8716–8723.

69 Duncan CJ, Smith JL, Greenaway HC: Failure to protect frog skeletal muscle from ionophore-induced damage by the use of the protease inhibitor leupeptin. Comp Biochem Physiol 1979;63C:205–206.

70 Weglicki WB: Degradation of phospholipids of myocardial membranes; in Wildenthal K (ed): Degradative Processes in Heart and Skeletal Muscle. Amsterdam, Elsevier/ North-Holland Biomedical Press, 1980;pp.377–388.

71 Katz AM: Membrane-derived lipids and the pathogenesis of ischaemiic myocardial damage. J Mol Cell Cardiol 1982;14:627–632.

72 Baracos V, Rodemann P, Dinarello CA, et al: Stimulation of muscle protein degradation and prostaglandin E_2 release by leucocytic pyrogen (Interleukin-1). N Engl J Med 1983;308:553–558.

73 Reeds PJ, Palmer RM: The possible involvement of prostaglandin $F_{2\alpha}$ in the stimulation of muscle protein synthesis by insulin. Biochem Biophys Res Commun 1983;116:1084–1090.

74 Jackson MJ, Wagenmakers AJM, Edwards RHT: Effect of inhibitors of arachidonic acid metabolism on efflux of intracellular enzymes from skeletal muscle following experimental damage. Biochem J, in press.

75 Jackson MJ, Edwards RHT: Free radicals, muscle damage and muscular dystrophy; in Quintanilha A (ed): Reactive Oxygen Species in Chemistry, Biology and Medicine. NATO, in press.

76 Jackson MJ, Jones DA, Edwards RHT: Vitamin E and muscle diseases. J Inherited Metab Dis 1985;8(suppl 1):84–87.

77 Morgan A, McDonald-Gibson RG, Slater TF: Arachidonic acid metabolism by rat uterus. Biochem Soc Trans 1984;12:837–838.

78 Mullane KM, Read N, Salmon JA, et al: Role of leukocytes in acute myocardial infarction in anaesthetised dogs: Relationship to myocardial salvage by anti-flammatory drugs. J Pharmacol Exp Ther 1984;228:510–522.

79 Grossman S, Waksman EG: New aspects of the inhibition of soyabean lipoxygenase by alpha-tocopherol. Evidence for the existence of a specific complex. Int J Biochem 1984;16:281–289.

80 Reddanna P, Rao MK, Reddy CC: Inhibition of 5-lipoxygenase by vitamin E. FEBS Lett 1985;193:3–43.

81 Godwin KO, Edwardly J, Fuss CN: Retention of ^{45}Ca in rats and lambs associated with the onset of nutritional muscular dystrophy. Aust J Biol Sci 1975;28:457–460.

82 Mokri B, Engel AC: Duchenne dystrophy: electron microscopic findings pointing to a basic or early abnormality in the plasma membrane of the muscle fibre. Neurology 1975;25:1111–1120.

83 Carpenter S, Karpati G: Duchenne muscular dystrophy, plasma membrane loss initiates muscle cell necrosis unless it is repaired. Brain 1979;102:147–161.

84 Bodenstein JB, Engel AG: Intracellular calcium accumulation in Duchenne dystrophy and other myopathies: a study of 567,000 muscle fibres in 114 biopsies. Neurology 1978;28:439–446.

85 Maunder-Sewry CA, Gorodetsky R, Yaron R, et al: Elemental analysis of skeletal muscle in Duchenne muscular dystrophy using X-ray fluorescence spectrometry. Muscle Nerve 1980;3:502–508.

86 Bertorini TE, Cornelio F, Bhattacharya SK, et al: Calcium and magnesium content in foetuses at risk and pre-necrotic Duchenne muscular dystrophy. Neurology 1985;34:1436–1440.

87   Jackson MH, Jones DA, Edwards RHT: Measurements of calcium and other elements in needle biopsy samples of muscle from patients with neuromuscular disorders. Clin Chim Acta 1985;147:215–221.
88   Tagesson C, Henrikkson KG: Elevated phospholipase A in Duchenne muscle. Muscle Nerve 1984;7:260–261.

M.J. Jackson, BSc, PhD, Department of Medicine, University of Liverpool,
PO Box 147, Liverpool, L69 3BX (UK)

7 Free Radicals, Central Nervous System Processes and Brain Functions

Walter B. Essman, Stuart B. Wollman

In the central nervous system a number of events can provide for the introduction of free radicals from oxygen. Radicals of the alkoxy and peroxy species, for example, can determine peroxidation reactions whereby the products serve as relevant structural and functional components of brain function. Oxygen-centered radicals, where the unpaired electron is chiefly localized to oxygen orbitals, are identified by electron spin resonance (ESR) as an anisotropic g tensor with an average value from 2.007 to 2.02. The products of oxygen radical peroxidation may couple with metal ions which, in turn, may increase the cytotoxicity of the free radicals.

In the brain there are numerous sources of oxygen-derived free radicals and these may exert a large variety of effects upon important central nervous system functions. For the most part, these effects are deleterious and, in some instances, have been suggested as model systems to account for important aspects of normal brain processes as well as those encountered in several clinicopathological entities. The purpose of this chapter is to consider some of the neural events with which oxygen-derived free radicals are associated, to examine the sources thereof, and to describe some of the agents or events that protect against the toxicity of free radicals.

One area of concern wherein oxygen-derived free radicals have been given a large measure of consideration is brain ischemia. Ischemic injury has been classically studied over a wide range of avenues of inquiry. In recent years increased attention has been directed toward the role of oxygen-derived free radicals.

Brain Ischemia

Events that provide for ischemia and cerebral tissue injury have been associated with free radicals. For example, vasospasm, which can certainly be a precursor of cerebral ischemic injury, has been shown to account for an increase in prostaglandin–indirectly an index of arachidonic acid formation through free radical production [1]. In cerebral atherosclerosis, another mechanism accounting for ischemic tissue injury, lipid peroxidation–suggestive of free radical activity–has been demonstrated [2]. During actual ischemic injury lipid peroxidation occurs [3]. The effects of anaerobic conditions differ from those of free radicals in cerebral ischemia [4]. The changes associated with ischemia-related peroxidative damage included altered glutathione redox couple, increased oxidized glutathione, and a degradation of polyenoic fatty acids and ethanolamine phosphoglycerides. In animal studies where the model for cerebral ischemia was occulsion of the middle cerebral artery in the cat, the changes produced included a reduction in the brain concentration of cholesterol, ascorbic acid, and polyenoic fatty acids [5–7].

Peroxidized lipid production in the brain as an index of ischemic injury has been accounted for by the production of electron stagnation in the mitochondrial electron transport system as a result of rapid oxygen deficiency [6].

The reduction of coenzyme Q by one unpaired electron from the electron transport chain then forms alkyl radicals. Hydrogen atoms are removed from unsaturated fatty acids of the inner mitochondrial membrane lipids. When alkyl radicals are increased, tissue oxygen is captured and there is autocatalytic lipid peroxidation.

It has been shown [7] that following embolization of the rat brain, weak chemiluminescence was demonstrated from the cortical surface. This well-known accompaniment of lipid peroxidation was indicative of the generation of an excited species from an early stage of reoxygenation around multiple ischemic areas. The free radicals implicated in the initiation of lipid peroxidation are singlet oxygen, the hydroxyl radical, and iron-oxygen and superoxide complexes. Ischemia-induced lipid peroxidation secondary to free radical generation is believed to bring about conformational changes in membrane receptor proteins, and consequently decreased ligand binding properties. Later data to be explored in this chapter will confirm that free radicals can alter the properties of receptors in the central nervous system.

The production of superoxide radicals at the cortical surface has been measured in cerebral vascular injury produced by experimentally induced hypertension [8]. On the basis of interaction and dismutation at the surface with superoxide dismutase (SOD), superoxide appears to occur in the extracellular space. Bases for this may depend upon a slow accumulation of leukocytes that contain superoxide (superoxide per cm^2 brain surface is equivalent to superoxide produced by one million activated leukocytes), vascular membrane damage by free radicals and/or their products, making superoxide more accessible to surface areas, or entry of superoxide into the extracellular space through membrane channels. The latter possibility was verified experimentally by showing that the irreversible anion channel inhibitor, phenylglyoxal, prevented superoxide formation from acute hypertension or from the topical application of bradykinin or arachidonic acid. Arachidonic acid can serve as a substrate for superoxide production, when acted upon by cyclooxygenase arachidonate in the presence of NADH. Inhibitors of cyclooxygenase, such as acetylsalicylic acid and indomethacin, inhibited the generation of superoxide radicals in this reaction. Certain specific prostaglandins, such as PGG_2, when used as substrates, were also capable of providing for superoxide generation. It has been suggested [9] that when acute increases in blood pressure occur, producing hyperemia, cerebral edema, and altered cerebral function, the bases for these effects may depend upon free radical-mediated processes.

The effects of free oxygen radicals on cerebral vessels results in alterations in the vascular smooth muscle and endothelium, relaxation of the vascular smooth muscle, and abnormal responses to vasomotor stimuli [10]. The effect, mainly upon the small arterioles, by arachidonic acid was attributed mainly to the effects of the free hydroxyl radical.

With cerebral anoxia or hypoxia associated with ischemia, irreversible neurological injury will occur with approximately 5 min of sustained anoxia [11, 12] due to interruption of the cerebral vascular circulation. The effects, in vitro, are less apparent; with complete ischemia of 20–30 min [13, 14] neurons just begin to show irreversible damage. If one were to accept the view that cellular damage following cerebral ischemia results from the formation of oxygen-derived free radicals, then further free radical generation and resulting cellular injury could occur with a reinstatement of oxygen, since tissue free fatty acids are elevated with transient hyperoxia [15]. Cerebral edema may occur from polyunsaturated fatty acids, and may be accounted for [16] by the transient formation of superoxide radicals. In concussive cerebral injury free radicals are generated and changes similar to those seen with vascular compromise are seen. In fact, pial arteriolar abnormalities are produced; this can be prevented by free-radical scavengers [17].

In the central nervous system ischemia and the reperfusion of ischemic tissue can produce mitochondrial damage [18]. Very shortly after the onset of cerebral ischemia, the maximum respiratory rate for phosphorylation is decreased in a graded manner, depending upon the duration of ischemia. Adenosine diphosphate (ADP)-stimulated respiration is inhibited by 50% after 15 min of ischemia and by 75% after 30 min of ischemia [19, 20].

A relationship between cerebral ischemia and damage to both neural and vascular tissue appears convincingly established. Such injury appears, in part, preventable, but not reversible by a variety of agents that interact directly or indirectly with either the free-radical generation process or the free radicals themselves. A separate section of this chapter will concern free radical protection. For example, as previously indicated, the ischemic injury associated with percussive cerebral injury has been associated with the generation of oxygen-derived free radicals. Both catalase and SOD applied topically were capable of protecting against sustained arteriolar dilatation, and reduced responsiveness to vasoconstrictor effects. Furthermore such topical treatment with superoxide scavengers attenuated the generation of superoxide in the extracellular space [21]. Lipid peroxidation in the ischemic brain of rats, produced by vascular occlusion, was shown to depend upon the generation of free radicals [22]. A peak in free radical spin adducts measured by ESR occurred for animals in which three vessel occlusion (bilateral carotids and basilar artery) was employed.

Ischemic brain injury (produced by occlusion of the middle cerebral artery in the cat) was prevented by treatment with 1, 2-bis-(nicotinamide) propane (AVS); cerebral edema was prevented. This effect has suggested [23] that free radicals and increased activity of hydroperoxidases in ischemia provide for stimulation of the microvascular Na^+-K^+-ATPase system, producing increased sodium flux across the blood-brain barrier. AVS acts as a scavenger of hydroxyl radicals, inhibits the stimulatory effect of a lipid hydroperoxide (15-HPAA) on the activity of Na^+-K^+-ATPase and the arachidonic acid cascade within the microvasculature. The mediation of ischemia-induced cerebral edema appears, therefore, to be mediated by the production of oxygen-derived free radicals.

The reperfusion of ischemic tissue produces changes that are also related to free radicals. Anoxia-induced depletion of adenosine triphosphate (ATP) in the brain could provide for the generation of superoxide during reperfusion. During ischemia there is an accumulation of intracellular free iron [24, 25]. There are several events that may account for such an increase in free iron; the liberation of iron from the mitochondria as a result of injury [26], the release of ferous iron from ferritin during reperfusion [27, 28]. The accumulation of ADP during ischemia can chealate Fe^{2+} and thereby provide for free radical formation during reperfusion [29, 30]. There is, therefore, further indication that iron can contribute to the cytoxic effects of free radicals produced during ischemia and with reperfusion of the ischemic brain.

There is a variety of other agents or events capable of either producing oxygen-derived free radicals, or augmenting free radical production by other conditions. The next section of this chapter will concern these circumstances.

Conditions for Free Radical Production in the Brain

Iron

In previous discussion, several studies have pointed to the role of iron in forming oxygen-derived free radicals. It has been pointed out that lipid peroxidation in the brain depends upon oxygen–and also upon the presence of iron ions [31, 32]. Certain regions of the brain contain unusually high concentrations of iron; e.g. the globus pallidus and substantia nigra [33–35]. Injury to these same brain regions appears specific to the effect of high-pressure oxygen [36]. When different regions of the rat brain were incubated under air or O_2, a 144% rate of lipid peroxidation was highest for the hypothalamus; with the addition of iron and ascorbic acid, lipid peroxidation was stimulated by 10-fold in the cerebral cortex and 20-fold in the hypothalamus. Lipid peroxidation under hyperbaric oxygen was inhibited when iron chelators were added–most markedly by α^2, α^1-dipyridyl, 1,10-phenanthroline, desferol, ethylenediaminetetraacetic acid (EDTA), and catechol [37]. Lipid peroxidation was not affected in this situation by catalase or peroxidase, suggesting that there may be a regional specificity, within the brain for the toxic effects of free radicals; this regional specificity may relate to the regional endogenous concentration of iron. Lipid peroxidation may be initiated by several ferrous chelate complexes, such as EDTA-Fe^{2+}, adenosine monophosphate (AMP)-Fe^{2+} and ADP-Fe^{2+} [38]. Lipid peroxidation reactions may occur, but only when Fe^{3+} is reduced to Fe^{2+}, in many instances as a result of the generation of superoxide as the reductant. Using brain synaptosomes, the initiation of lipid peroxidation was controlled by the ratio of Fe^{3+} to Fe^{2+} [39]. Within the range of 1:1 to 7:1, the iron ratios appeared optimal for peroxidation reactions. It has been suggested that lipid peroxidation reactions that are produced by chelators, oxidizers or reducing agents may ultimately depend upon the resulting ratio of Fe^{3+} to Fe^{2+}. Some further support for this possibility may be found in studies where lipid peroxidation initiated by ethanol was regulated, in rats, by the extent of iron loading [40], thus again raising the possibility that at optimal Fe^{3+}–Fe^{2+} ratios the peroxidation of membrane lipids can be augmented, whereas at ratios that exceed 7:1, peroxidation reactions may be attenuated by iron.

Another primary vehicle for iron delivery to the central nervous system is hemoglobin. Studies in the effects of purified hemoglobin on brain homogenates and spinal fluid of anesthetized cats showed a marked inhibition of the activity of Na-K-ATPase and the initiation of lipid peroxidation. Both of these effects of hemoglobin were blocked by the iron chelator, desferoxamine [41]. Since desferoxamine chelates only free iron ions and not heme iron, the proposed effects of hemoglobin are likely to be accounted for by free iron ions.

Hyperoxia
The effects of hypoxia have been reviewed earlier, particularly as related to ischemia. Aside from the generation of oxygen-derived free

radicals and the peroxidation of neural lipids, there are pathological changes observed. In adult rats there is edema of the myelinated tracts, spongy neuropil, dark neurons and neuronolysis, while in neonatal rats, hypoxia causes a more marked spongy neuropil [42]. With hyperoxia more marked pathological changes occur. In the newborn rat, moderately severe spongy neuropil occurs with severe karyorrhexis of the neurons. Among young adult rats, hyperoxia causes moderately severe spongy neuropil and neuronolysis. In neonatal rats exposed neonatally to 3 h of hyperoxia, a very rapid rise in cerebral free fatty acid levels occurred, suggesting lipid peroxidation of neural membranes [43] – probably induced by oxygen-derived free radicals. Direct oxygen toxicity has been described in brain histologically [44] and the effects of free radicals upon both neurons and glia have been described [45, 46]. The maturation of the brain and vascularization of the developing nervous system can be affected by hyperoxia [47, 48] – changes which again are consistent with free radical effects.

Calcium

Another factor which contributes to oxygen-derived free radical action in the central nervous system is calcium. It has been argued that shifts in calcium across the neuronal membrane provide for an intracellular calcium overload, a cascade of events leading to neuronal damage and loss [15, 49]. When free radicals are generated in neural tissue, there is a calcium-mediated injury to mitochondria [18] and the release, from the mitochondria, of the mitochondrial iron pool [26]. These findings would certainly agree with those previously discussed relating iron to free radical effects. Free radical-mediated cell damage can partially depend upon intracellular calcium [50]. Synergistic effects of calcium and free radicals have been demonstrated in mitochondrial membranes from rat brain and upon mouse spinal cord [51]. An increased intracellular calcium appears to account for this effect.

Acidosis

The effects of acidosis upon free radical formation and lipid peroxidation have been assessed [52]. In rat cortical tissue, moderate to marked acidosis (pH = 6.5–6.0) grossly increased free radical production and increased the degradation of phospholipid-bound polyenoic fatty acids. When the pH was further reduced to 5.0, the effects were reversed, suggesting that moderate to severe acidosis (6.5–6.0) represents a pH optimum for free radical formation. As previously considered, the effects of ischemia, in producing acidosis, may actually promote free radical formation within this reduced pH range. This effect may be accounted for by an increased formation of the protonated form of the superoxide radical; this form is strongly prooxidant and lipid-soluble, and may alter iron binding to macromolecules.

Prostaglandin

Oxygen radicals involved in the synthesis of prostaglandins have been previously considered as having an important role in vascular and tissue injury in the nervous system attending acute hypertensive episodes. After brain injury there is an increase in the concentration of prostaglandin in tissue [53]. The topical application of arachidonic acid or bradykinin induced an increased synthesis of prostaglandin and an increased production of superoxide radicals [54]. Such free radical formation was inhibited by the topical application of SOD. A very marked reduction in superoxide production was effected by topical application of $4,4^1$-diisothiocyano-2-2^1-stilbene disulfonate and phenylglyoxal; these compounds are anion channel inhibitors. These findings suggest that where plostaglandin synthesis in the brain is increased, there is an increased metabolism of arachidonic acid by cyclooxygenase, and superoxide radicals appear in the extracellular space. The anion channel permits the traversal of superoxide ions across the membranes of intact cells.

Radiation

Ionizing radiation is well appreciated as a source of cellular disruption. The brain is also sensitive to the effects of γ radiation. Mice were exposed to ionizing radiation and lipid extracts of brain tissue analyzed with electron paramagnetic spectroscopy (EPR) indicated the formation of radicals; these could have derived from molecular interaction with radiation or from the reaction of water radiolysis products with cell components at the membrane interface [55]. Spectral studies of lipid extracts from several organelles after radiation revealed an oxygen-centered lipid radical confined to the plasma membrane-nuclear fraction. Such findings could suggest the organelle specificity of ionization radiation effects upon the brain.

Adriamycin

The in vivo administration of adriamycin to rats brought about a change in hydroxyl radicals, measured in brain tissue from reaction products that such radicals form with salicylates. As compared with untreated controls, adriamycin-treated rats had a 55% increase in hydroxyl radical formation [56]. It is clear from these results that hydroxyl radicals are formed after adriamycin treatment and this effect, aside from occurring in other tissues, is very apparent in the brain.

Methyl-4-Phenyl-1,2,5,6-Tetrahydropyrridine

The toxicity of methyl-4-phenyl-1,2,5,6-tetrahydropyrridine (MPTP) upon the dopamine-containing neurons of the brain has been demonstrated and verified [57–59]. It has been appreciated that MPTP is oxidized after its entry into the brain and the toxic product, 1-methyl-4-phenyl pyrudinum (MPP) is formed. The latter is catalyzed by monoamine oxidase type B (MAO-B) outside of dopamine-containing cells, probably in glia [60–62]. A transport system specific for the dopaminergic neuron can carry the toxic MPP to the dopaminergic neurons where it is taken up [62, 63]. It has been suggested that the intraneuronal toxicity of MPP can be accounted for by excess free radicals [64, 65]. The peroxidation of dopamine, which itself constitutes a basis for free radical generation, can be potentiated by MPTP and MPP [65].

In studies where mice were injected with MPTP, this caused a marked decrease in intraneuronal dopamine; this effect was not attenuated by antioxidants, suggesting that the neurotoxicity of MPTP is mediated by free radicals either inaccessible to antioxidants (compartmentalized) or that the free radical initiation of the neurotoxic effect is extracellular and the intraneuronal effects are due to other causes [66]. There can be little doubt that free radicals are responsible, if not directly then indirectly, for the neurotoxicity of MPTP.

Aging

It has been suggested that changes in the brain that occur with aging are mediated by cellular damage caused by free radical reactions [67]. The view that aging and free radical activity are correlated in the brain has been advocated [68, 69]. This view suggests that brain aging depends upon the interaction of oxygen radicals with iron ions. This interaction depends upon a rich supply of brain lipids, a relatively low level of scavenger enzyme activity in the brain, and a high rate of oxygen utilization.

Oxygen free radicals in the brain account for lipid peroxidation, particularly phospholipids rich in polyunsaturated fatty acids [70]. Highly reactive carbonyl fragments, particularly malondialdehyde, are liberated when lipid peroxides decompose; malondialdehyde may produce cross-linking of amine-containing compounds, resulting in Schiff base materials; this appears to be the basis for fluorescent lipofuscin [71, 72].

Thiobarbituric acid-reactive products are formed in the brain as a direct result of malondialdehyde action [73], so that thiobarbituric acid-reactive products may provide an index of brain changes associated with aging. Regional differences in brain thiobarbituric acid-reactive products demonstrated when adult rat brains were compared with aged rat brain [74]. Significant increases in thiobarbituric acid-reactive products were observed in aged rats for the cerebral cortex, septal area, hippocampus, caudate, putamen, and substantia nigra. SOD activity was decreased with aging in the frontal cortex, septal area, caudate putamen, and substantia nigra.

The role of aging-related events in the production of oxygen-derived free radicals in the brain is further supported by studies of age pigments. Two such pigments, ceroid and lipofuscin, are deposited as insoluble polymeric complexes in cells and subcellular organelles as a consequence of metal catalyzed peroxide decomposition. Metallic ions were shown to stimulate autoxidation of lipids in phospholipid membranes and fatty acid micelles [75]. This finding appears to be consistent with the observation that important constituents of the age pigments are iron and copper [76]. It appears as if the formation of age pigments in the brain, notably in aging, depends upon the presence of catalytic metal complexes. Further evidence for a relationship between age pigments and free radicals derives from studies in which there was an inversely proportional relationship between lipofuscin content in the Torpedo central nervous system and the activity of SOD [77]. The data support the view that age pigments are products of free lipoperoxidation by free radicals. The decrease in SOD activity observed in some studies [78] serves as a basis for increased superoxide ion available for lipid peroxidation reactions and also serves as a means by which oxygen-derived free radical action may be induced. There was a substantial (25% decrease in SOD activity in the substantia nigra of aged rats [74]).

Aluminum

The deposition of aluminium in brain cells has been an associated finding in several entities including dialysis dementia [79], Guam Parkinson dementia, Alzheimer's disease [80, 81], and amytrophic lateral sclerosis and related demyelinating disorders [82]. In neurons showing structural changes, e.g. neurofibrillary degeneration, the excess aluminum appears to be associated with the nucleus [83, 84]. Several bases for the localization of aluminium to the nucleus may be found; these include: (a) aluminum blocks the initiation sites for RNA polymerase [85]; (b) binding of aluminum to DNA phosphate and bases [86]; (c) aluminum increases histone DNA binding, (d) blocks sister chromatin exchange [87] in isolated rabbit brain nuclei, physiological concentrations of aluminium lactate. A 50–60% inhibition of ADP-ribosylation activity [88]; this inhibitory effect persists for at least 20 days after aluminum lactate injection thereby correlating well with the pathological and behavioral effects [80, 83].

Electroconvulsive Shock

Numerous ongoing chemical and physical processes in the brain are altered by the seizure activity induced by a single electroconvulsive shock. This treatment has been used experimentally to model seizure disorders and to serve as an analog for the treatment used to treat depressive illness. As an example of how regional synaptosomes are capable of manifesting

Cerebral cortex (16.4 µg/mg)

Cerebellum (6.8 µg/mg)

Diencephalon (14.1 µg/mg)

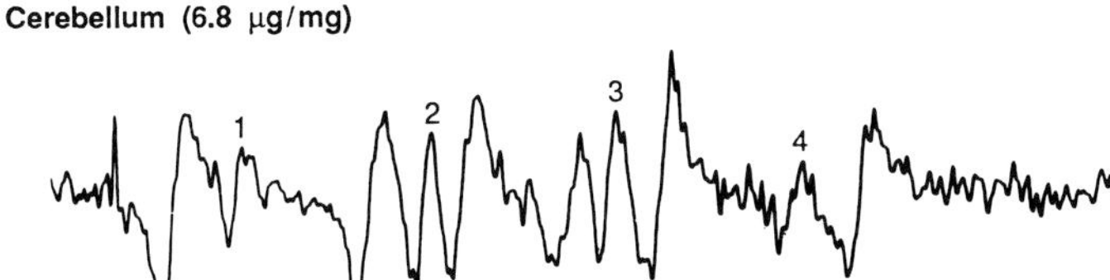

Fig. 1. The baseline EPR spectra for free radical spin adducts are shown, using synaptosomal membrane fractions derived from the mouse cerebral cortex, diencephalon, and cerebellum. Only some notable hydroxyl radicals may be identified for the diencephalon, but not for membranes deriving from the other brain regions.

oxygen-derived free radicals, fig. 1 shows a baseline EPR spectrum for synaptosomes derived from three major regions of the mouse brain. In figures 2, 3, and 4 baseline levels are again shown, in contrast with regional synaptosomal spectra based upon a single in vivo electroconvulsive shock. It is apparent that in synaptosomes from each region, greater oxygen-derived free radical production occurs in the synaptosomal region as a consequence of electroconvulsive shock and its seizure sequelae.

There is the important question of aluminum involvement in the formation of brain free radicals. Some experimental findings that deal with this relationship have been summarized in tables 1–3.

Now that a number of means by which oxygen-derived free radicals in the brain and related interactions have been examined and a number of interrelationships and interdependencies have been explored, it is apparent that such free radicals exert a multitude of effects in the brain. It is the purpose of the next section to detail some of these effects.

Effects of Free Radicals upon Brain Functions

Perhaps one of the most relevant aspects of the issue of oxygen-derived free radicals in the brain is the impact of such radicals upon a variety of cellular, tissue, and functional processes within the nervous system.

One relevant compartment of the brain affected by free radicals is the blood-brain barrier; this physicochemical barrier maintains control over the permeability of a large number of molecules from blood stream into the brain substance. When O_2-derived free radicals were generated through the application of arachidonate to the brain surface of cats, the permeability to proteins was increased [89]. Both albumin and horseradish peroxidase were increased in their cortical deposition following topical arachidonate. The increase in blood-brain barrier permeability produced by O_2-derived free radicals occurred mainly in penetrating arterioles, but not in pial arterioles or veins. It is quite reasonable that brain capillaries are affected directly by free radicals; since the capillary endothelial cell is rich in xanthine oxidase, a ready source of free radicals exists. Microvessels, when isolated from rat brain, were studied when in a free-radical generating system [90]. There was a 74% reduction in capillary uptake of rubidium, whereas efflux processes appeared to be unaffected. Such an effect would be consistent with a decrease in Na^+-K^+-ATPase activity in the endothelial cell, resulting from free radical production. A direct relationship between superoxide radicals and rat brain-derived Na^+-K^+-ATPase has been described [91]; in the presence of superoxide radicals rat brain membrane-derived Na^+-K^+-ATPase was irreversibly inactivated.

In brain mitochondria oxygen-derived free radicals interfere with mitochondrial respiration [24]. Mitochondria isolated from rat brain were subjected to a free radical generating system and oxygen consumption by the isolated organelles was measured. Significant inhibition of stage 3 respiration was produced. Since the inhibitory effect was not modified by the addition of SOD, the effect was largely attributable to the action of the hydroxyl radical.

Fig. 2–4. The effect of a single in vivo electroconvulsive shock (ECS) resulting in a full clonic-tonic convulsion, prior to obtaining and processing brain tissue from mice, has been investigated for free radical formation. In figures 2–4 changes in oxy-radical spin adducts are shown for the cerebral cortex, diencephalon, and cerebellum, respectively. Baseline activity in each case is changed, so that a single ECS appears clearly to generate oxygen-derived free radicals, as measured by ESR spectroscopy.

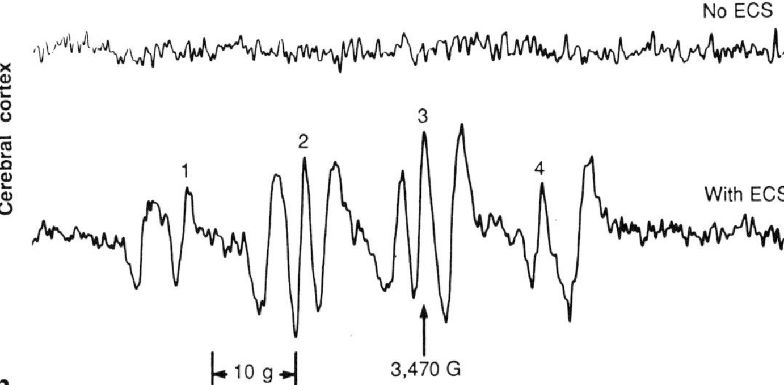
No ECS
Cerebral cortex
1
2
3
4
With ECS
10 g
3,470 G
2

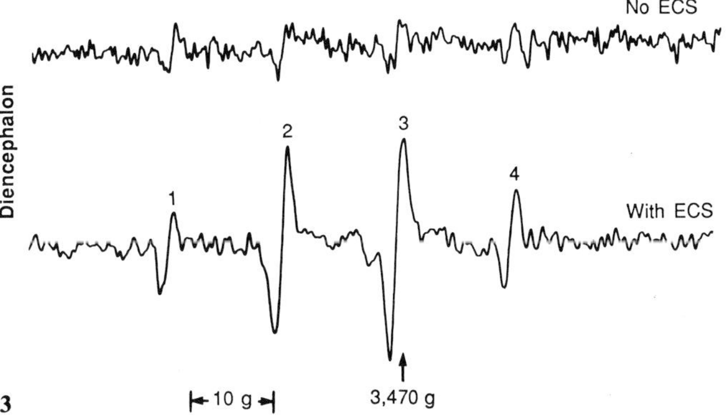
No ECS
Diencephalon
1
2
3
4
With ECS
10 g
3,470 g
3

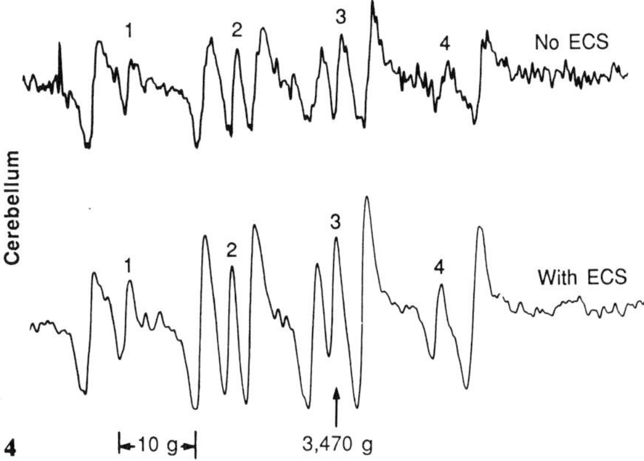
1
2
3
4
No ECS
Cerebellum
1
2
3
4
With ECS
10 g
3,470 g
4

Table 1. Levels of aluminum in regions of the mouse brain following aluminum gluconate treatment

Age months	Brain region					
	cerebral cortex	hippocampus	basal ganglia	limbic area	brain stem	cerebellum
6	1.6 (0.3)	3.8 (0.4)	2.4 (0.5)	0.6 (0.1)	0.3 (0.1)	0.6 (0.1)
12	1.9 (0.6)	3.9 (0.2)	2.8 (0.4)	0.5 (0.1)	0.4 (0.1)	0.7 (0.2)
18	2.2 (0.6)	4.6 (0.8)	3.6 (0.8)	0.6 (0.2)	0.5 (0.1)	0.8 (0.1)
24	3.1 (0.9)	5.2 (1.3)	4.2 (1.1)	0.7 (0.3)	0.5 (0.2)	0.5 (0.1)
30	3.4 (1.1)	6.2 (1.5)	5.4 (1.3)	0.6 (0.1)	0.4 (0.1)	0.3 (0.1)

Concentrations in μg/g. All values are expressed as mean ($\pm$ SD) for 10 mice at each age.

Table 2. Subcellular distribution of aluminum in the hippocampus of mice at different ages

Age months	Subcellular fraction			
	cytoplasm	nucleus	mitochondria	synaptosome
6	18 (4)	126 (17)	12 (3)	89 (12)
12	10 (2)	138 (24)	8 (2)	93 (18)
18	11 (2)	268 (19)	15 (3)	136 (15)
24	20 (3)	487 (27)	12 (2)	233 (35)

Concentrations in μg/g. Values = mean $\pm$ SD.

Table 3. Transformed peak second derivatives for hippocampal tissue from mice of several ages

Age months	Treatment	
	saline	aluminum gluconate
6	0.0	4.2
12	0.2	3.9
18	1.3	6.7
24	2.7	9.3

All observations are based upon 10 samples for each condition.

Oxygen-derived free radicals have already been indicated as altering vascular permeability, transport processes and enzyme activity. The effects of free radicals upon the responses of cerebral vessels resemble, in many respects, the effects produced by ischemia, severe acute intracranial hypertension and repeated seizures [10]. Such effects include vasodilation and reduced responsiveness. Further support for the view that vascular damage in the brain is caused by the hydroxyl radical derives from studies [92] showing that cerebral arterioles in cats showed sustained dilatation, became unresponsive to the vasoconstrictor effects of hypocapnia, and showed destructive lesions of the endothelium and vascular smooth muscle. The electrical resistance of brain vascular endothelium was studied in frogs [93]. Topically generated free radicals caused a decreased resistance of the cortical endothelium within 1–2 s. These observations also support the view that oxygen-derived free radicals may mediate ischemic damage to cortical endothelium.

The synaptic effects of free radicals are relevant for a number of important functional events that depend upon the integrity of the synapse and its substructure. Synaptosomal Na^+-K^+-ATPase activity was inhibited by the production of free radicals [94]. A dose-related inhibitory effect was produced by free radicals. In hippocampal brain slices from the guinea pig, hydroxyl radicals decreased synaptic efficacy [95]. Somatic and dendritic responses to orthodromic stimulation were affected. Such effects would again be consistent with the inhibition of Na^+-K^+-ATPase–in this instance, synaptically based.

The age-related increase in oxygen-derived free radicals are associated with neurotransmitter-related changes in the brain [96]. Decreases in choline acetyltransferase, and acetylcholinesterase; MAO and γ-aminobutyric acid (GABA) transaminase activity were increased. Free radicals generated from membrane-unsaturated fatty acids can modulate GABA binding to its receptor on cerebral cortical synaptic membranes from rats [97]. The uptake of GABA by synaptosomes isolated from rat brain was inhibited by free radical generation [98]. It was determined that the hydroxyl radical specifically mediated this effect. The inhibition of synaptosomal GABA uptake by generation of hydroxyl radicals could be blocked by the pretreatment with corticosteroids. The prevention of uptake inhibition of GABA by steroids is assumed to depend upon a block in lipid peroxide formation. In response to free radicals generated by peroxidation, rat brain synaptosomes, the high-affinity uptake of several neurotransmitters was significantly modified. GABA and choline uptake were reduced whereas, by 5 min, dopamine uptake was appreciably increased. There was also a suppression of calcium uptake by synaptosomes [99]. The translocation of synaptosomal molecules by free radicals appears related to effects upon the synaptosomal membrane. Nonlipid damage from free radicals may relate to effects upon cation transport proteins.

An oxidative mechanism specific to the damage of dopamine-containing neurons in the nigrostriatum has been described [100]. The neurotoxicity of such dopamine neuron-specific agents as 6-hydroxydopamine and 6-aminodopamine may be explained on the basis of hydrogen peroxide-derived superoxide radicals. The neurotoxicity of 5,7-dihydroxytryptamine is mediated by MAO and the resulting oxygen-derived free radicals. In the human brain, the substantia nigra has an appreciably lower concentration of reduced glutathione, as compared with other brain regions. In autopsied brain tissue from patients with Parkinson's disease, reduced glutathione was completely absent in the striatum [101]. Reactive oxygen species generated by the oxidative deamination of L-dopa and dopamine may be responsible for damage to membranes and other subcellular organelles. A nigrostriatal deficiency in reduced glutathione could render this region increasingly susceptible to oxidative injury; thereby a potential basis for the etiology of Parkinson's disease may be suggested by these findings.

Modification of Free Radical Action in the Brain

There is a large number of agents and events which are capable of either blocking or attenuating either the production or effects of oxygen-derived free radicals in the brain. Such effects have had relevance for previous discussion. Such effects may represent a number of mechanisms of action including: (a) replacement of an oxidatively deaminated substrate; (b) blocking phosphorylation; (c) producing a membrane-fixing or coating effect; (d) enzyme inhibition; (e) scavenger enzyme action; (f) increasing scavenger enzyme activity; (g) antioxidants; (h) metalic ion chelation; (i) block peroxidation, and (j) alteration of ion channels. Although all of these mechanisms have been suggested by experiments involving cells, subcellular organelles, tissue, or brain regions.

It remains for further experimental investigation to elucidate those mechanisms underlying the formation of oxygen-derived free radicals in the brain. The role of free radicals in the mediation of effects of drugs on the central nervous system, as well as for pathology and functional alteration within the brain, appears to become increasingly clear. With further clarification of such relationships the etiology and potential treatment of such conditions may become accessible.

References

1   Sasaki T, Murota S-J, Wakai S, et al: Evaluation of prostaglandin biosynthetic activity in canine basilar artery following subarachnoid injection of blood. J Neurosurg 1981; 55:771–778.

2   Wilson RB, Middleton CC, Sun GY: Vitamin E antioxidants and lipid peroxidation of experimental athero-sclerosis of rabbits. J Nutr 1978; 108:1858–1867.

3   Demopoulos H, Flamm E, Seligman M, et al: Molecular pathology of lipids in CNS membranes; in Jobsis FF (ed): Oxygen and Physiological Function. Dallas, Professional Information Library, 1977, pp 491–508.

4   Rehncrona S, Smith DS, Åkesson B, et al: Peroxidative changes in brain cortical fatty acids and phospholipids, as characterized during Fe^{2+}-and ascorbic acid-stimulated lipid peroxidation in vitro. J Neurochem 1980; 34:1630–1638.

5   Demopoulos HB, Flamm ES, Pietronigro DD, et al: The free radical pathology and the microcirculation in the major central nervous system disorders. Acta Physiol Scand 1980;492(suppl):91–119 (1980).

6   Flamm ES, Demopoulos HB, Seligman HL, et al: Possible molecular mechanisms of barbiturate-mediated protection in regional cerebral ischemia. Acta Neurol Scand 1977; 56(suppl 64): 150–151 (1977).

7   Flamm ES, Demopoulos HB, Seligman HL, et al: A free radicals in cerebral ischemia. Stroke 1978; 9:445–447.

8   Kogure K, Arai H, Abe K, et al: Free radical damage of the brain following ischemia. Prog Brain Res 1985; 63:237–259.

9   Kontos HA: Oxygen radicals in cerebral vascular injury. Circ Res 1985; 57:508–516.

10   Kontos HA, Wei EP, Christman CW, et al: Free oxygen radicals in cerebral vascular responses. Physiologist 1983; 26:165–169.

11   Weinberger IM, Gibbon MH, Gibbon JH: Temporary arrest of circulation to the central nervous system. I. Physiologic effects. Arch Neurol Psych. 1940; 43:615–620.

12   Heymans C: Survival and revival of nervous tissue after arrest of the circulation. Physiol Rev 1950; 30:375–392.

13   Ames A III: Earliest irreversible changes during ischemia. Am J Emerg Med 1983; 1:139–146.

14   Hossman KA: Neuronal survival and revival during and after cerebral ischemia. Am J Emerg Med 1983; 1:191–197.

15   Siesjo BK: Cell damage in the brain: a speculative synthesis. J Cereb Blood Flow Metabol 1981; 1:155–185.

16   Chan PH, Fishman RA: Transient formation of superoxide radicals in polyunsaturated fatty acid-induced brain swelling. J Neurochem 1980; 35:1004–1007.

17   Wei EP, Kontos HA, Dietrich WD, et al: Inhibition by free radical scavengers and by cyclooxygenase inhibitors of pial arteriolar abnormalities from concussive brain injury in cats. Circ Res 1981; 48:95–103.

18   Fiskum G: Involvement of mitochondria in ischemic cell injury and in regulation of intracellular calcium. Am J Emerg Med 1983; 1:147–153.

19   Hillered L, Ernster L, Siesjo BK: Influence of in vitro lactic acidosis and hypercapnia on respiratory activity of isolated rat brain mitochondria. J Cereb Blood Flow Metab 1984; 4:430–437.

20   Hamud F, Fiskum G: Loss of maximal respiratory and Ca^{2+} uptake capacities by rat brain mitochondria during cerebral ischemia (abstract). Biophys J 1985; 47:414a.

21   Kontos HA, Wei EP: Superoxide production in experimental brain injury. J Neurosurg 1986; 64:803–808.

22 Tominaga T, Imaizumi S, Yoshimoto T, et al: Detection of free radicals generated in NADPH-dependent lipid peroxidation in ischemic brain homogenate–application of spin trapping technique. No To Shinkei 1986; 38:169–175.

23 Asano T, Johshita H, Gotoh O, et al: The pathomechanism underlying ischemic brain edema: the role of Na^+-K^+-ATPase of the brain microvessel. No Shinkei Geka Neurolog Surg 1985; 13:1147–1159.

24 Hillered L, Ernster L: Respiratory activity of isolated rat brain mitochondria following in vitro exposure to oxygen radicals. J Cereb Blood Flow Metabol 1983; 3:207–214.

25 Gutteridge JMC, Rowley DA, Halliwell B: Superoxide dependent formation of hydroxyl radicals in the presence of iron salts. Detection of 'free' iron in biological systems by using Adriamycin-dependent degradation of DNA. Biochem J 1981; 199:263-265.

26 Tangeras A, Flatmark T, Backstrom D, et al: Mitochondrial iron not bound in heme and iron sulfur centers: Estimation, compartmentation, and redox state. Biochim Biophys Acta 1980; 589:162–175.

27 Mazur A, Green S, Saha S, et al: Mechanism of release of ferritin iron in vivo by xanthine oxidase. J Clin Invest 1958; 37:1809–1817.

28 Jones T, Spencer R, Walsh C: Mechanism and kinetics of iron release from ferritin by dihydroflavins and analogues. Biochemistry 1978; 17:4011–4017.

29 Tien M, Svingen BA, Aust SD: Initiation of lipid peroxidation by perferryl complexes; in Rogers MAJ, Powers EL, (eds): Oxygen and Oxy-Radicals in Chemistry and Biology, New York, Academic Press, 1981, pp 147–152.

30 Aust SD, Svingen BA: The role of iron in enzymatic lipid peroxidation. Free Radicals Biol 1982; 5:1–28.

31 Barber AA: Lipid peroxidation in rat tissue homogenates: Interaction of iron and ascorbic acid as the normal catalytic mechanism. Lipids 1966; 1:146–151.

32 Vladimirov YUA, Olenev VI, Suslova TB, et al: Lipid peroxidation in mitochondrial membrane. Adv Lipid Res. 1980; 17:173–249.

33 Hallgren B, Sourander P: The effect of age on the non-haemin iron in the human brain. J Neurochem 1958; 3:41–51.

34 Harrison WW, Nestsky MG, Brown MD: Trace elements in human brain: copper, zinc, iron and magnesium. Clin Chim Acta 1968; 21:55–60.

35 Francois C, Nguyen-Legros J, Percheron G: Topographical and cytological localization of iron in rat and monkey brains. Brain Res 1981; 215:317–322.

36 Balentine JD, Gutsche BB: Central nervous system lesions in rats exposed to oxygen at high pressure. Am J Pathol 1966; 48:107–127.

37 Zaleska MM, Floyd RA: Regional lipid peroxidation in rat brain in vitro: possible role of endogenous iron. Neurochem Res 1985; 10:397–410.

38 Bucher JR, Tien M, Aust SD: The requirement for ferric in the initiation of lipid peroxidation by chelated ferrous iron. Biochem Biophys Res Commun 1983; 111:777–784.

39 Braughler JM, Duncan LA, Chase RL: The involvement of iron in lipid peroxidation. Importance of ferric to ferrous ratios in initiation. J Biol Chem 1986; 261:10282–10289.

40 Nordmann R, Ribiere C, Rouach H: Involvement of iron and iron-catalyzed free radical production in ethanol metabolism and toxicity. Enzyme 1987; 37:57–69.

41 Sadrzadeh SM, Anderson DK, Panter SS, et al: Hemoglobin potentiates central nervous system damage. J Clin Invest 1987; 79:662–664.

42 Ahdab-Barmada M, Moossy J, Nemoto EM, et al: Hyperoxia produces neuronal necrosis in the rat. J Neuropathol Exp Neurol 1986; 45:233–246.

43 Lin MR, Barmada MA, Nemoto EM: Detrimental cerebral metabolic effects of normobaric oxygen in one-day-old rats (abstract). Anesthesiology 1983; 59(suppl 3A):A398.

44 Detrisac C, Greene WB, Balentine JD: Ultrastructural pathology of central nervous system oxygen toxicity, in vitro (abstract). Lab Invest 1982; 46:18A–19A.

45 Chan PH, Yurko M, Fishman RA: Phospholipid degradation and cellular edema induced by free radicals in brain cortical slices. J Neurochem 1982; 38:525–531.

46 Chan PH, Fishman RA: Alterations of membrane integrity and cellular constituents by arachidonic acid in neuroblastoma and glioma cells. Brain Res 1982; 248:151–157.

47 Sokoloff LB: Toxic effects of elevated oxygen tension on brain maturation; in Berenberg SR (ed): Brain. Fetal and Infant. Current Research on Normal and Abnormal Development. Den Haag, Nijhoff, 1977, pp 178–183.

48 Hannah RS, Hannah KJ: Hyperoxia: Effects on the vascularization of the developing central nervous system. Acta Neuropathol 1980; 51:141–144.

49 Hass WK: Beyond cerebral, metabolism and ischemic thresholds: An examination of the role of Ca^{2+} in the initiation of cerebral infarction; in Meyer JS, Lechner H, Reivich M, et al(eds): Cerebral Vascular Disease. Proc Saltzburg Conf on Cerebral Vascular Disease. Amsterdam, Excerpta Medica, 1981, vol 3, pp. 3–7.

50 Casini AF, Farber JL: Dependence of the carbon tetrachloride induced death of cultured hepatocytes on the extracellular calcium concentration. Am J Pathol 1981; 105:138–148.

51 Braughler JM, Duncan LA, Goodman T: Calcium enhances in vitro free radical-induced damage to brain synaptosomes, mitochondria, and cultured spinal cord neurons. J Neurochem 1985; 45:1288–1293.

52 Siesjo BK, Bendek G, Koide T, et al: Influence of acidosis on lipid peroxidation in brain tissues in vitro. J Cereb Blood Flow Metab 1985; 5:253–258.

53 Ellis EF, Wright KF, Wei EP, et al: Cyclooxygenase products of arachidonic acid metabolism in cat cerebral cortex after experimental concussive brain injury. J Neurochem 1981; 37:892–896.

54 Kontos HA, Wei EP, Ellis EF, et al: Appearance of superoxide anion radical in cerebral extracellular space during increased protaglandin synthesis in cats. Circ Res 1985; 57:142–151.

55 Lai EK, Crossley C, Sridhar R, et al: In vivo spin trapping of free radicals generated in brain, spleen, and liver during γ radiation of mice. Arch Biochem Biophys 1986; 244:156–160.

56 Floyd RA, Henderson R, Watson JJ, et al: Use of salicylate with high pressure liquid chromatography and electrochemical detection (LCED) as a sensitive measure of hydroxyl free radicals in adriamycin treated rats. J Free Radicals Biol Med 1986; 2:13–18.

57 Burns RS, Chiueh CC, Markey SP, et al: A primate model of parkinsonism: selective destruction of dopaminergic neurons in the pars compacta of the substantia nigra by N-methyl-4-phenyl-1,2,3,6-tetrahydropyridine. Proc Natl Acad Sci USA 1983; 80:4546–4550.

58 Hallman H, Lange J, Olson L, et al: Neurochemical and histochemical characterization of neurotoxic effects of 1-methyl-4-phenyl-1,2,3,6-tetrahydropyridine on brain catecholamine neurones in the mouse. J Neurochem 1985; 44:117–127.

59 Heikkila RE, Manzino L, Cabbat FS, et al: Protection against the dopaminergic neurotoxicity of 1-methyl-4-phenyl-1,2,5,6-tetrahydropyridine by monoamine oxidase inhibitors. Nature 1984; 311:467–469.

60 Chiba K, Trevor A, Castagnoli N Jr: Metabolism of the neurotoxic tertiary amine, MPTP, by brain monoamine oxidase. Biochem Biophys Res Commun 1984; 120:574–578.

61 Langston JR, Irwin I, Langston EB, et al: 1-Methyl-4-phenylpyridinium ion MPP$^+$: identification of a metabolite of MPTP, a toxin selective to the substantia nigra. Neurosci Lett 1984; 48:89–92.

62 Melamed E, Rosenthal J, Globus M, et al: Mesolimbic dopaminergic neurons are not spared by MPTP neurotoxicity in mice. Eur J Pharmacol 1985; 114:97–100.

63 Javitch JA, Snyder SG: Uptake of MPP$^+$ by dopamine neurons explains selectivity of parkinsonism-inducing neurotoxin, MPTP. Eur J Pharmacol 1985; 106:455–466.

64 Mytilineou C, Cohen G: Deprenyl protects dopamine neurons from neurotoxic effect of 1-methyl-4-phenylpyridinium ion. J Neurochem 1985; 45:1951–1953.

65 Poirier J, Donaldson J, Barbeau A: The specific vulnerability of the substantia nigra to MPTP is related to the presence of transition metals. Biochem Biophys Res Commun 1985; 128:25–33.

66 Martinovits G, Melamed E, Cohen O, et al: Systemic administration of antioxidants does not protect mice against the dopaminergic neurotoxicity of 1-methyl-4-phenyl-1,2,5,6-tetrahydropyridine (MPTP). Neurosci Lett 1986; 69:192–197.

67 Harman D: The aging process. Proc Natl Acad Sci USA 1981; 78:7124–7128.

68 Harman D: Free radical theory of aging: consequences of mitochondrial aging. Age 1983; 6:86–94.

69 Floyd RA, Zaleska MM, Harmon J: Possible involvement of iron and oxygen free radicals in aspects of aging in brain; in Armstrong D, Sohal RS, Cutler RG, Slater RF (eds): Free Radicals in Human Biology, Aging and Disease. New York, Raven Press, 1984, pp 143–161.

70 Fong K-L, McCay PB, Poyer JL, et al: Mesolimbic dopaminergic neurons are not spared by MPTP neurotoxicity in mice. Eur J Pharmacol 1985; 114:97–100.

71 Tappel AL: Lipid peroxidation damage to cell membrane. Fed Proc 1973;32:1870–1874.

72 Tappel AL: Measurement of and protection from in vivo lipid peroxidation; in Pryor WA (ed): Free Radicals in Biology. New York, Academic Press, 1980, vol 4, pp 1–47.

73 Gutteridge JMC: Free-radical damage to lipids, amino acids, carbohydrates and nucleic acids determined by thiobarbituric acid reactivity. Int J Biochem 1982; 14:649–653.

74 Mizuno Y, Ohta K: Regional distributions of thiobarbituric acid-reactive products, activities of enzymes regulating the metabolism of oxygen free radicals, and some of the related enzymes in adult and aged rat brains. J Neurochem 1986; 46:1344–1352.

75 Gutteridge JMC: Age pigments: role of iron and copper salts in the formation of fluorescent lipid complexes. Mech Ageing Dev 1984; 25:205–214.

76 Siakotos AN, Goebel HH, Patel V, et al: The morphogenesis and biochemical characteristics of ceroid isolated from cases of neuronal ceroid-lipofuscinosis; in Volk BW, Aranson SM (eds): Sphingolipids, Sphingolipidoses and Allied Disorders. New York, Plenum Press, 1972, pp 53–61.

77 Totaro EA, Pisanti FA: Stereological analysis of lipofuscin in the central nervous system of *Torpedo marmorata*: correlation with superoxide dismutase distribution. Experientia 1985; 41:1047–1048.

78 Ono T, Okada S: Unique increase of superoxide dismutase level in brains of long living mammals. Exp Gerontol 1984; 19:349–354.

79 Alfrey AC, LeGendrie GR, Kaehny WD: The dialysis encephalopathy syndrome: possible aluminum intoxication. N Engl J Med 1976; 294:184–188.

80 Crapper McLachlan DRT, Krishnan SS, Dalton AJ: Brain aluminum distribution in Alzheimer's disease and experimental encephalopathy. Science 1973; 180:511–513.

81 Crapper McLachlan DRT, Krishnan SS, Quittkat S: Aluminum neurofibrillary degeneration and Alzheimer's disease. Brain 1976; 99:67–80.

82 Yase Y: The role of aluminum in CNS degeneration with interaction of calcium. Neurotoxicology 1980; 1:101–110.

83 Petit TL, Biederman GB, McMullen P: Neurofibrillary degeneration, dendritic dying back and learning-memory deficits following aluminium administration: implications for brain aging. Exp Neurol 1980; 67:152–162.

84 Perl DP, Brody AR: Intraneuronal aluminum accumulation in amyotrophic lateral sclerosis and Parkinson-dementia of guam. Science 1982; 217:1053–1055.

85 Sarkander HI, Balb G, Schlosser R, et al: Blockade of neuronal brain RNA initiation sites by aluminum: a primary molecular mechanism of aluminum-induced neurofibrillary changes? in Cervos-Navarro J, Sarkander HI (eds): Brain Aging: Neuropathology and Neuropharmacology. New York, Raven Press, 1983, vol 21, pp 259–274.

86 Karlik SJ, Eichron GL, Lewis PN, et al: Interaction of aluminum species with deoxyribonucleic acid. Biochemistry 1980; 19:5991–5998.

87 DeBoni U, Seger M, Crapper McLachlan DRT: Functional consequences of chromatin bound aluminum in cultured human cells. Neurotoxicolocy 1980; 1:65–81.

88 Crapper McLachlan DRT, Dam T-V, Farnell BJ, et al: Aluminum inhibition of ADP-ribosylation in vivo and in vitro. Neurobehav Toxicol Teratol 1983; 5:645–647.

89 Wei EP, Ellison MD, Kontos HA, et al: O_2 radicals in arachidonate-induced increased blood-brain barrier permeability to proteins. Am J Physiol 1986; 251:H693–H699.

90 Lo WD, Betz AL: Oxygen free-radical reduction of brain capillary rubidium uptake. J Neurochem 1986; 46:394–398.

91 Hexum TD, Fried R: Effects of superoxide radicals on transport (Na + K) adenosine triphosphatase and protection by superoxide dismutase. Neurochem Res 1979; 4:73–82.

92 Kontos HA, Hess ML: Oxygen radicals and vascular damage; in Spitzer JJ (ed): Myocardial Injury. New York, Plenum Press, 1983, vol 161.

93 Olesen S-P: Free oxygen radicals decrease electrical resistance of microvascular endothelium in brain. Acta Physiol Scand 1987; 129:181–187.

94 Koide T, Asano T, Matsushita H, et al: Enhancement of ATPase activity by a lipid peroxide of arachidonic acid in rat brain microvessels. J Neurochem 1986; 46:235–242.

95 Pellmar T: Electrophysiological correlates of peroxide damage in guinea pig hippocampus in vitro. Brain Res 1986; 364:377–381.

96 Noda Y, McGeer PL, McGeer EG: Lipid peroxides in brain during aging and vitamin E deficiency: possible relations to changes in neurotransmitter indices. Neurobiol Aging 1982; 3:173–178.

97 Yoneda Y, Kuriyama K, Takahashi M: Modulation of synaptic GABA receptor binding by membrane phospholipids: possible role of active oxygen radicals. Brain Res 1985; 33:111–122.

98 Braughler JM: Lipid peroxidation-induced inhibition of γ-aminobutyric acid uptake in rat brain synaptosomes: protection by glucocorticoids. J Neurochem 1985; 44:1282–1288.

99   Dabrowiecki Z, Gordon-Majszak W, Łazarewicz J: Effects of lipid peroxidation of neurotransmitters uptake by rat synaptosomes. Pol J Pharmacol Pharm 1985; 37:325–331.
100 Cohen G: Monoamine oxidase, hydrogen peroxide, and Parkinson's disease; in Yahr MD, Bergmann KJ (eds): Advances in Neurology. New York, Raven Press, 1986, vol 45, pp 119–125.
101 Perry TL, Godin DV, Hansen S: Parkinson's disease: A disorder due to nigral glutathione deficiency? Neurosci Lett 1982; 33:305–310.

Walter B. Essman, MD, PhD, Department of Psychology, Queens College,
City University of New York, 65–30 Kissena Boulevard,
Flushing, NY 11367–0904 (USA)